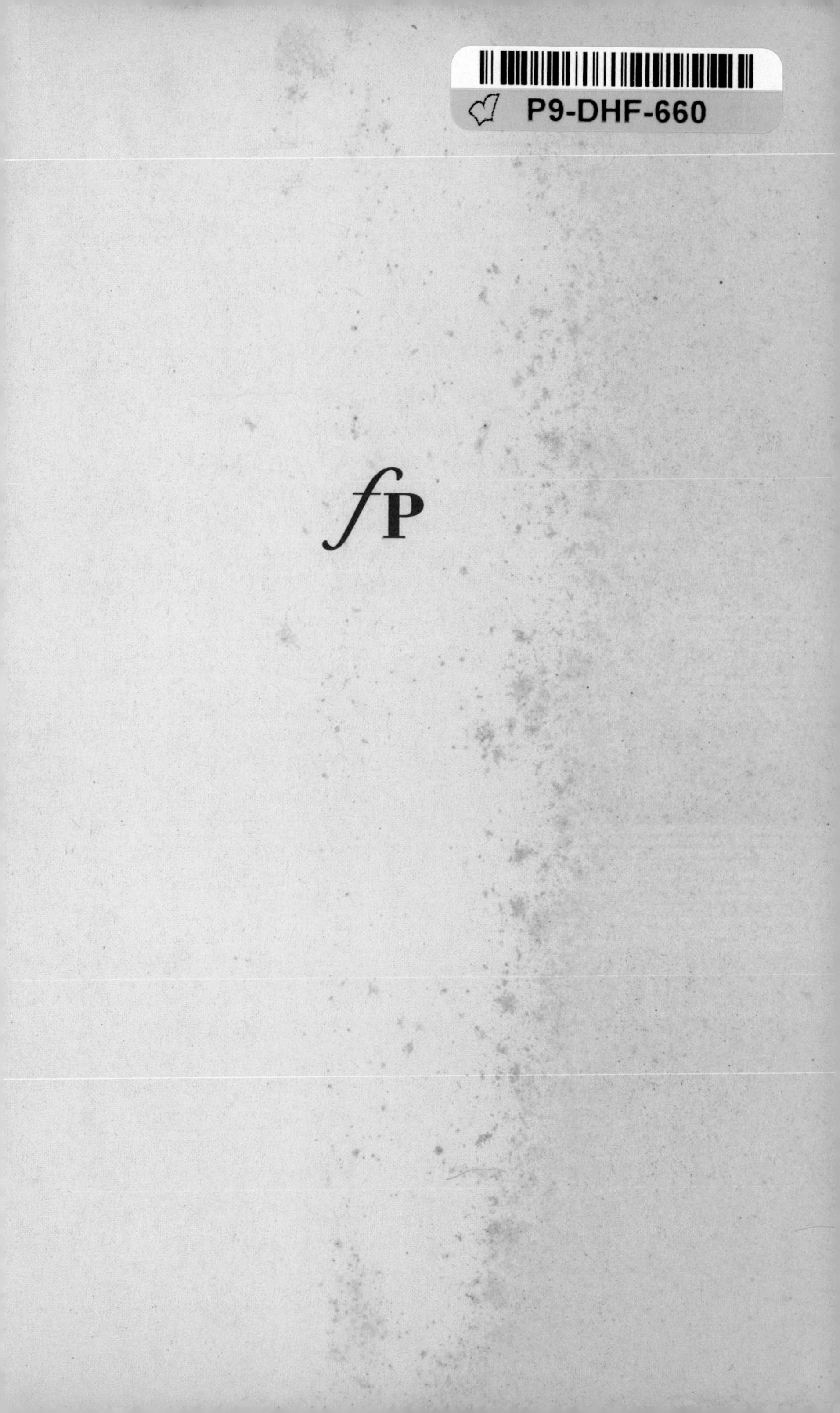

P9-DHF-660
fP

Also by Stanley Coren

The Intelligence of Dogs
How Dogs Think
What Do Dogs Know? (with Janet Walker)
Why We Love the Dogs We Do
How to Speak Dog
The Pawprints of History
Sleep Thieves
The Left-Hander Syndrome

Why Does My Dog Act That Way?

A Complete Guide to Your Dog's Personality

Stanley Coren

Free Press
New York London Toronto Sydney

fP
FREE PRESS
A Division of Simon & Schuster, Inc.
1230 Avenue of the Americas
New York, NY 10020

First Free Press trade paperback edition December 2007

FREE PRESS and colophon are trademarks of Simon & Schuster, Inc.

For information about special discounts for bulk purchases, please contact Simon & Schuster Special Sales at 1-800-456-6798 or business@simonandschuster.com

Designed by Davina Mock

Manufactured in the United States of America

1 3 5 7 9 10 8 6 4 2

The Library of Congress has catalogued the hardcover edition as follows:
Coren, Stanley.
Why does my dog act that way? A complete guide to your dog's personality/Stanley Coren.
p. cm.
Includes bibliographical references and index.
1. Dogs—Behavior. 2. Dogs—Psychology. I. Title.
SF433 .C694 2006
636.7—dc22 2006050966

ISBN-13: 978-0-7432-7706-8
ISBN-10: 0-7432-7706-6
ISBN-13: 978-0-7432-7707-5 (pbk)
ISBN-10: 0-7432-7707-4 (pbk)

This book is dedicated to my father, Benjamin Coren,
and to a little dog named Spirit who keeps him company.

CONTENTS

PREFACE

As a psychologist, I have always wondered why some of the techniques that we use to predict human behaviors are not used on dogs, particularly that part of psychological science that deals with personality. I believe that each dog has a unique, measurable personality—a product of the dog's genetic makeup and life history. As with humans, once you know your dog's personality, you can reasonably predict its behaviors in many circumstances and also recognize why your dog may behave differently from another dog in the same situation.

One special feature of this book is that it will show you how to create a *Superdog,* not one who flies and wears a cape, but a dog that is friendly, fearless, cooperative, intelligent, and trainable. The procedures are derived from the latest scientific data and involve actually shaping your dog's brain through particular handling and rearing practices and by deliberately arranging aspects of his living conditions. Some of this information is derived from documents describing fascinating research done by the U.S. Army Biosensor program, which have only recently been released under the U.S. Freedom of Information Act.

In this volume you will find a lot of new information, including an extensive listing of the personality profiles of 133 breeds of dogs, which I compiled based on data from 96 dog experts. I thank those experts for the many hours that they spent providing this material. This book also contains the first scientific study of heroic and helping behaviors in dogs based on 1,006 documented reports.

Finally, this book will provide you with a simple procedure to measure the personality of your own dog and compare it to other dogs of its breed. The test is valid and can be applied regardless of whether your dog is purebred or not. You can use the test results to predict your dog's behavior in many situations, and this can allow you to anticipate certain kinds of problems to which your dog may be susceptible.

Along with all this science and information, you will encounter stories that describe the interesting, exciting, and often unexpected behaviors of dogs. You will meet dogs that are heroes, helpers, friends, and loyal family members as well as a few unsavory examples of the species. You will develop a better understanding of the pet at your feet and see why he or she is neither a wolf in sheep's clothing nor a four-footed human in a fur coat.

On a personal note, I would also like to thank my wife, Joan, who went through the first draft of this book and offered many useful comments as well as some that are not printable. The book is much better (and shorter) because of her contribution.

Why Does My Dog Act That Way?

CHAPTER ONE

Personality and Temperament: Predicting What Lassie Will Do Next

According to certain Chinese legends, Buddha summoned all of the animals to him when he was ready to leave the earth. His purpose was to say good-bye and give the world some final advice. Only twelve animals came: the rat, ox, tiger, rabbit, dragon, snake, horse, sheep, monkey, rooster, pig, and of course, the dog. To acknowledge the special loyalty of these creatures above the others, Buddha named a year after each animal.

The Chinese say that if you are born in a given animal's year then you share its personality. Or, more elegantly, "This is the animal that hides in your heart." Every twelfth year is the year of the dog, and people born in 1934, 1946, 1958, 1970, 1982, 1994, and 2006 are all supposed to have dog-like personalities. Supposedly, people born in the year of the dog are honest, loyal, friendly, and protective. Such people are also stubborn. According to this belief, dogs, and people born under their sign, tend to worry too much, don't like crowds or noisy parties, can sometimes be snappish, and often don't like to involve themselves in the disputes of others.

The charm of the Chinese astrological system is that most people can agree that nonhuman animals have a distinctive character, similar to what we would call personality in humans. We even use animal labels to describe the personality characteristics of other humans—for example, we might call them "pigs," "sharks," "weasels," "catty," or "a lamb." We are also quite comfortable when someone says something like, "Elvis is really sociable, friendly, and extroverted," even though we know that she is referring to her golden retriever and not to the "King of Rock and Roll" or any other human being. Furthermore, we know that she means to tell us that we can approach this dog, safely touch it, that he will not shy away from us, will wag his tail, and seem happy to meet us. In her mind she is not only predicting but also explaining her dog's behavior. Elvis will act friendly because he has a friendly and sociable nature or personality.

The typical dictionary definition of personality is "the totality of an individual's behavioral and emotional characteristics." Psychologists study personality because they believe that understanding the personality of an individual allows us to predict how that person will act, react, and feel in various situations. Thus, the science of personality attempts to dissect, label, and catalog the typical ways that specific humans think, behave, and display their emotions in order to explain and predict why people act the way they do. To give a simple example, a person described as having an "aggressive personality" would be expected to act in a more hostile manner in most circumstances than would someone described as having a "warm and loving personality."

Some researchers think personality can only be useful in describing humans. Some psychologists, biologists, and animal behaviorists get very uncomfortable when we speak about a dog's personality. They would say that such comments tell us more about the person who is describing the dog than about the dog itself. They question whether a dog's mental structure is similar enough to a human being's for us to be able to analyze and describe the animal's behaviors with the same labels and methods that we use for people.

Personality and Temperament

The most fundamental use of the concept of personality is to allow us to predict the future behaviors of a person. Obviously, we would like to predict how a dog will act in various situations as well. So if we find general, systematic trends in the behaviors of individual dogs, and these allow us accurately to predict their behaviors in different circumstances over a large portion of each dog's life span, then it would seem sensible to talk about canine personality.

One reason some behavioral scientists don't want to accept that animals can have personalities similar to those of people is because this can stir up complex issues having to do with the way the mind works. Some theorists find it hard to talk about a human being's personality without eventually having to deal with questions about consciousness, responsibility, and even morality. Although some personality theories only focus on common behaviors and how they can be predicted, some others are very philosophical in their nature. Some of these "humanistic theories" involve concepts of an "ideal self," notions of art, beauty, justice, humanity, and the "greater social good." Obviously, if such concepts were included in the idea of personality, then it would not be sensible to apply the term to animals.

Scientists who feel uncomfortable using the word "personality" to describe animals are apt to use the term *temperament* instead when speaking about behavior of other species. Although the average person will see little difference between temperament and personality, to a psychologist these labels mean different things. Temperament typically refers to something more primitive than personality. It focuses on basic aspects of behavior, such as an individual's activity level and irritability, so it can be examined in very simple living things. Even an earthworm has a describable activity level and a measurable irritability, although no one seriously thinks that an earthworm has higher-level mental processes. When dealing with more-complex animals, such as dogs and young human infants, the concept of temperament is normally expanded to include emotional predispositions of various types. Thus, an individual might be described as fearful, aggressive, easily soothed, or calm, but once again matters such as conscious thought and judgment are apt to be avoided. Using a different label allows the theorist to suggest that there are still significant qualitative differences between the behaviors of people and animals.

New research, however, suggests that these theorists are probably being far too restrictive and that the same methods of analysis that allow us to measure the personality of humans seem to work for dogs. Knowing the personality of a dog may actually help us to understand, explain, and predict its behavior better. For these reasons, I will risk annoying my scientific colleagues and will use the words temperament and personality as if they mean the same thing.

Spaniels Are from Venus and Terriers from Mars

The Chinese astrological system assumes that all dogs have a similar personality, but everybody knows that among more earthly bound canines the personality of a dog varies widely based on its breed. This gives rise to statements like:

—Beagles don't really ever grow up. They simply skip the whole aging process and go from one form of puppyhood to another.

—Golden retrievers are not dogs—they are wet kisses on paws.

—The Airedale terrier believes that nothing is of any use to anyone unless it provokes a furor.

—Rottweilers spend most of their time thinking angrily about nothing.

—Each year a healthy Jack Russell terrier consumes one and a half times his weight in human patience.

—Boxers act as if there is a perfectly valid objection to every possible human request.

—A Newfoundland often enrages smaller dogs by demonstrating their inability to enrage him.

—Fox terriers have the ability to compress the largest amount of activity into the smallest number of thoughts.

—Bulldogs display that typically English characteristic for which there is no English name.

—All poodles act as if they have won first prize in the lottery of life.

—The beautiful and elegant Afghan hound knows two things: first that they are not very smart and, second, that it doesn't matter.

—All spaniels have a way of getting to the answer "yes" without ever having posed any clear question.
—The Chihuahua's greatest ambition is to live in a hot country and watch its master throw stones in the sea.
—A Labrador retriever's idea of guarding the house involves falling asleep in front of the doorway so that a burglar will trip over him in the dark and make enough noise to waken the rest of the family.
—Pekingese do not have owners—they have servants.
—One may expect that intensive training will have the same effect on basset hounds as it has on river rocks.
—Border collies are expensive to keep because they do not reach their full potential unless they are allowed to graduate from an Ivy League university.

While these are meant to be amusing descriptions of different dog breeds, they do seem to resonate with most people, probably because each contains an element of truth about the different personalities of the breeds, or at least fits the popular stereotypes of how such dogs behave.

Most people pay careful attention to the breed of the dog that they bring into their home because they have the feeling that particular breeds have characteristics that can cause problems or promote happiness. For some people certain breeds simply "don't work" while others "are perfect." The critical factor determining the success of the relationship seems to be the match, or mismatch between the dog's personality and the human's personality and lifestyle. Sometimes the concerns are wider and include matching the personality of a new dog with that of another dog or cat already living in the house.

It certainly is the case that most dog and person mismatches are quite breed specific, as shown by a series of surveys conducted by the American Animal Hospital Association. They estimate that up to one-third of all dogs acquired by people are subsequently voluntarily given up to shelters and pounds, euthanized, or simply abandoned. In most instances the problem is that the dog's behavior just doesn't fit in with the person's lifestyle and personality. Of the people who give dogs up to a shelter, it is estimated that over 40 percent have had successful relationships with other breeds of dogs. About half of the people who give up their dog to a shelter go on to get another dog, with 93 percent of these choosing another breed. About 90 percent of these "second marriages" seem to be successful, suggesting that the problem is most likely due to the fact that another dog, with a different personality, was a better choice.

Reagan and Rex

One example of the importance of a dog's personality comes from former U.S. president Ronald Reagan. Reagan liked dogs and had many during his life. Before his political career started he had a pair of Scottish terriers (one was named Scotch and the other Soda). Later, as governor of California and then as president of the United States, he received many dogs as gifts. Among these was a golden retriever named Victory, an Irish setter named Peggy, a Siberian husky named Taca, and a Belgian sheepdog named Fuzzy. Of these, Reagan had favorites. For example, he always spoke particularly fondly of his Scottish terriers.

Shortly after Reagan arrived at the White House a Bouvier des Flandres was given to his wife, Nancy, and it was named Lucky after Mrs. Reagan's mother, Edith Luckett Davis. Although given to Nancy, it was the president who usually took responsibility for the family's dogs, and Lucky's temperament did not mesh well with his. Bouviers are large dogs that have been specialized to herd cattle. Although quite friendly, they can be dominant and demanding. Such a pushy dog just didn't fit in with the president's daily routine or his concept of the dignity of his office. Lucky became an embarrassment when Reagan entertained the British prime minister Margaret Thatcher. The press carried photographs of Lucky dragging the president around the White House lawn at the end of a leash much to the amusement of his visitor. On another occasion, Lucky let her herding instinct go to her head and nipped at the presidential hindquarters, a trick Bouviers use to make cattle move more quickly. Unfortunately, this time she nipped hard enough to draw blood. This incident was also caught by a press photographer, which caused further embarrassment to the president. Reagan did not have the patience or the personality traits required to interact successfully with this rambunctious dog and Lucky was ultimately "retired" to the Reagan ranch in Santa Barbara, California, thus removing her annoying presence from the president's daily life.

Lucky was replaced by a Cavalier King Charles spaniel named Rex. The Reagans first encountered the breed on a visit to publisher and press commentator William F. Buckley. Nancy liked the look of these little dogs, but Ronald was captivated by their friendly personality. Reagan arranged to get one of these dogs from Buckley (who kept Rex's littermate, Fred), ostensibly as a Christmas gift for Nancy, but it soon became clear that the little dog belonged to the president, at least emotionally. Rex patrolled the West Wing corridors looking for Reagan and could often be found guarding the Oval Office door when staff and visitors were escorted in to see the president. If possible he would sneak into the office to curl at the president's feet. Sometimes he even tried to jump into his lap—which was difficult for such a small dog to do without assistance—while at other times he managed to climb onto a chair or sofa to be next to his master. Reagan would pretend not to notice, but would occasionally let his hand wander over to pet the little white-and-chestnut-colored dog. Although not noted for frivolous sentimentality, Reagan had a fancy doghouse built for Rex. It was designed by Theo Hayes, great-great-grandson of President Rutherford Hayes, and had red window draperies and framed pictures of the president and first lady as interior decorations.

Shortly after leaving the White House, Reagan's health began to deteriorate. Because of Ronald's advancing Alzheimer's disease, he and Nancy moved from Rancho del Cielo to Bel Air, and, of course, took Rex with them. Rex spent much of his time with Reagan, resting beside him as he had done in the Oval Office. Although Reagan's memory weakened to the point where he recognized very few people, he did seem to recognize Rex and would often call him by name when he was not nearby. Thus the little dog whose personality still seemed to fit with that of the now frail Ronald Reagan stayed near him for hours on end. Rex died six years before his master and was sorely missed.

Darwin's Dog

Lucky's personality clash with her master exiled her from the White House, but she did live out the rest of her life comfortably on the Reagan ranch. For other dogs, a personality clash can have a much darker outcome. Take as an example a case involving Charles Darwin, whose theory of evolution changed the nature of our understanding of the biological world. Darwin was always fond of dogs and as a young man had a way with them. He even caused a bit of a family spat when he proved that his affinity with dogs was strong enough to allow him to steal the affection of a dog that had been given to his sister. Later, when a university student at Cambridge, he won the love of his cousin's dog, who would creep down inside his bed and sleep by his feet each night.

Darwin was a keen observer of dogs, and often used his experiences with his own dogs as examples of the similarity between human and animal behavior. In his book *The Expression of the Emotions in Man and Animals,* dog behavior is featured prominently.

Darwin liked dogs that were both active and affectionate. When his children were young his favorite was a large black and white retriever cross named Bob. This dog was well loved by his children and Darwin would spend long hours watching his family and the dog interact. Bob taught Darwin so much about loyalty, empathy, and affection in dogs that he ended up using the dog's behavior as examples of these qualities in some of his books.

Darwin's all-time favorite dog came to him late in his life. She was a rough white fox terrier named Polly who would be mentioned many times in his books, but only by the description "my white terrier," and never by name. Darwin and Polly had a strong bond. She had a special basket where she would curl up and watch him work in his study. After a morning of scholarly reading and writing, regardless of whether the weather was wet or clear, Darwin would go out for a long walk. Polly went with him in fair weather, but when it rained, she balked. Darwin's son noted that she "might be seen hesitating in the veranda, with a mixed expression of disgust and shame at her own want of courage; generally, however, her conscience carried the day, and as soon as he was evidently gone she could not bear to stay behind."

Polly was active, sharp witted, and affectionate, and Darwin delighted in telling people how she would know that her master was going away on a journey because she noticed the signs of packing going on around her. He was flattered by the fact that she seemed distressed at his leaving. She would sulk for days in his absence, but when Polly noticed that her master's study was being prepared for his return, she would become quite excited, and frequently check the door or look out of the window in anticipation of his coming home.

There was an amusing flow of communication between Darwin and Polly. She had developed the strategy of trembling and putting on an air of misery when Darwin passed her on the way to the kitchen or dining room. She seemed to know that he would respond to her dramatic act by announcing that "she is famishing," as he often did, and reward her with a treat "lest she die of hunger." She did have to work a bit for this extra snack, since Darwin made her catch biscuits that he first placed on the end of her nose, preceded by an affectionate and mock-solemn lecture that she must "be a very good girl."

Unfortunately, Darwin was once given a dog that was the exact opposite of Polly in every aspect of her personality, and hence was fated not to work its way into the naturalist's affection. The dog was a Talbot hound and was given to Darwin by a friend who thought that he might be interested in the breed, because of its history and genetic contribution to other, more popular, dog breeds. The breed was first brought to England after William, Duke of Normandy, defeated Harold, King of England, in the Battle of Hastings on October 14, 1066. These predominantly white hounds were favorites of William, and had been introduced to him by his councilor and ally Roger de Montgomery, who was an avid breeder of Talbot hounds. Roger came to England following William's victory, and received extensive grants of land in different parts of the kingdom, eventually becoming the Earl of Shrewsbury. Sometime later the Montgomery family had a falling-out with one of William's successors, King Henry I, and in the mid-1400s the title of Earl of Shrewsbury, along with the estates and the breeding kennels, were given to John Talbot. The new earl kept the kennels active and ultimately the Talbot family developed a deep fondness for the dogs that had been so carefully bred by their predecessor. In fact, the Talbots went so far as to name the breed after their own family. They even changed their family coat of arms to contain a white hound above the shield to signify their association with this line of dogs.

Relatively slow moving, with large ears and resonant voice, Talbot hounds became part of the foundation of our modern bloodhounds, although modern bloodhounds are somewhat more active and much more affectionate. Talbot hounds were also bred with smaller dogs and are believed to have contributed their fine scenting and tracking abilities to the quick-moving beagle. As a breed, Talbot hounds are relatively inactive dogs, preferring to spend much of the day sleeping in the sun. Talbots seem to be only occasionally stirred to movement, and this tends to occur when an interesting scent reaches their exquisitely sensitive noses, at which point they will follow it loudly announcing their find with a deep baying sound. They are typically disinterested in learning obedience commands (which is common to a number of hound breeds), and this makes them difficult to train. Furthermore, although they can be friendly when people initiate interactions, they make little effort on their own to be sociable and affectionate.

Darwin's Talbot hound (whose name was never recorded) swiftly became an irritant to the great scientist. The problem was clearly the dog's personality. Darwin described it as "graceless, noisy, and drooling," "witless and lacking in self-control," and "with no visible merit of consequence to civilized society." In the end, at a time when the children were not around to be disturbed by the events, the terrier-loving Darwin had the hapless Talbot hound taken from the house and disposed of.

It thus becomes clear that a dog's personality certainly has an effect on his relationship with people since it predicts the behaviors the dog is most likely to produce. These behaviors then can have a profound influence on the positive or negative outcome of his life. Even in situations in which you might expect that the most important factor in a dog's success would be his intelligence and trainability, personality plays an important role. Many service dogs, such as guide dogs for the blind, or drug- and bomb-detecting dogs, fail their training programs—not because they are not smart enough to do the work—but rather because of personality flaws, such as fearfulness or aggressive tendencies.

A Collie with Personality

I am reminded of a statement made by Rudd Weatherwax, the dog trainer who gave us Lassie, which illustrates the importance of canine personality. Of all of the dogs that have ever graced the movie screen or starred on television, Lassie has been the most popular. Lassie's career began with the release of the film *Lassie Come Home* in 1943 starring Roddy McDowall, Elizabeth Taylor, and a dog named Pal who played the role of Lassie. In 1954 the first Lassie television show appeared and eventually the character of Lassie went through seven different TV families. The franchise has been showing on the television screen now for more than fifty years, and there have been nine generations of Lassie, as of this writing (all of whom are descendents of the original Pal). Rudd's son Robert Weatherwax eventually took over the training and careers of the new generations of Lassies and continues it today. Lassie is so popular and recognizable that the dog was even called upon to be an Academy Award presenter in 1982, for which he received a standing ovation from a theater full of other movie stars.

In 1950, Lassie was being billed as "the smartest dog in the world." Weatherwax could boast that Lassie knew over ninety commands. "He can string these commands together the way that we string words together to make sentences. Just like each sentence has a different meaning, we can put together different series of commands to make up unique routines in front of the camera."

Obviously, Pal was an incredibly intelligent, well-trained dog. With that in mind, one might expect that, when asked what made Lassie different from other dogs, Rudd Weatherwax would say it is Pal's intelligence and trainability that make him special. Instead, he answered, "I like to explain it as personality. It is his personality, a dog's personality, that has made him a star in motion pictures and given him a fan following that is, in numbers and in loyalty, probably unsurpassed by that of any other dog in history . . . Hollywood has now found the warmth of a beautiful collie dog."

If personality is what makes Lassie a special dog, that alone might motivate us to want to study the personality of dogs. However, understanding a dog's personality also has practical uses. A dog owner who knows his dog's personality can predict how his dog will act and react even in situations that the dog has never encountered before. Canine personality also informs us as to which dogs are best for certain tasks, occupations, and lifestyles. In fact, the original process of domesticating dogs was actually an attempt to change a wild animal into one that had an acceptable personality that would make it a useful and loving companion for humans.

CHAPTER TWO

Creating Dogs

Several experts in animal behavior have told me that "Dogs are just wolves in sheep's clothing." Because they believe that dogs are simply domesticated wolves, they use the extensive body of scientific data collected from observing the behavior of wolves in the wild as a shortcut to understanding dog behavior. Unfortunately, this ignores the fact that dogs and wolves are different species. No one disputes the fact that dogs and wolves evolved from a common ancestor or that they share enough instinctive behaviors from that ancestry to make comparisons interesting and useful. But the ways in which their behaviors differ are many and perhaps are even more interesting.

Trying to predict the behavior of one animal species based upon the behavior of another species is speculative at best—even if these species have a substantial degree of genetic overlap. For example, few scientists interested in social or political behavior would argue that we ought to study relationships among the members of a chimpanzee troop because it would give us insight into human behavior—such as why humans prefer leaders with particular personal characteristics. Yet humans are nearly as close genetically to chimpanzees as dogs are to wolves.

Some researchers have argued that dogs are an "invented" species because humans created dogs by manipulating the characteristics of wild canines until they had something that we now call our domestic dog. In doing so, humans not only altered the physical characteristics of dogs, so that they no longer look like wolves, but also there is every reason to believe that humans also manipulated canine behaviors and personalities so that today's dogs no long act like wolves or other wild canines.

A Successful Species

In evolutionary terms, dogs have been a lot more successful than wolves. Evolution measures success by population numbers; a species is considered to be successful if it produces many offspring and those offspring go on to thrive and reproduce. There is no worldwide census to determine the size of the dog population, and getting precise numbers would be difficult in any event, since in many countries people don't keep dogs inside their homes as pets. In some places, the dogs simply roam freely in the streets and nobody really owns them, so counting would be quite difficult. Nevertheless, the best scientific guess is that there are more than 400 million dogs in the world. To get an idea of how many dogs this is, we would have to add the total number of people in the United States, Canada, Great Britain, and France to get as many humans as there are dogs in the world! In contrast, best scientific data available suggests that if we add together all of the wolves from all of the countries in the world they would number only about 400,000. That means that there are a thousand times more dogs in the world than wolves.

Part of the success of dogs has to do with their breeding strategies. Domestic dogs usually can breed twice a year, while wolves and other wild canines breed only once a year. In addition, dogs are quite promiscuous and will breed at any opportunity. In most wolf packs there is only a single breeding pair, which annually produces one litter, usually of three to six pups. The first litter does not usually appear until a female is two years old. The puppies are in constant danger from predators—often including humans who often treat wolves as an unwanted and potentially harmful species, which puts both the pups and the adults upon which the pups depend at risk.

Domestication changes animal breeding patterns. Compare the breeding pattern of dogs to those of wolves: a female dog can have her first litter of puppies when she is only five to eighteen months old (depending upon her breed). It takes fifty-eight to seventy days to have the puppies. The average number of pups in each litter is usually between six and ten. Every female dog can have two batches of puppies each year. Now, if half of these puppies are females, they obviously can also have pups when they mature. That means that one female dog and her offspring could produce 4,372 puppies in seven years! The survival rate of the puppies will also be a lot higher in domestic dogs, since they will usually have human assistance to protect and care for them.

The real difference between dogs and wolves, however, is not simply their breeding patterns or their dramatically varied physical shapes and sizes. Neither a Yorkshire terrier nor an English mastiff, for instance, looks at all like a wolf. The wee terrier is smaller than any wolf, while the giant mastiff is larger than any wolf alive today. Yet these striking physical differences are not what make dogs different from wolves. Rather, it is their temperament and personality which were manipulated by humans during the process of domestication.

The First Mother of All Dogs

We may never know the complete and unambiguous story about how dogs came to be domesticated. We are, however, beginning to get some idea of when and where domestication took place. The trail of the early dog is faint, and part of it is written in the genetic makeup of the species. Recent studies compared the DNA of dogs to the DNA of 39 different wild species of canids, including wolves, jackals, dingoes, and coyotes. The DNA that geneticists use for these studies is not the material that we think of as genes, which is found in the chromosomes in the nucleus of cells. Half of that nuclear DNA comes from the mother and half from the father. The nuclear DNA is unique to each individual because the genetic material from each parent is a chance selection of half of each parent's chromosomes. This makes the DNA of the offspring unique for each mating. However, the DNA that helps us trace the origin of dogs is not found in the nucleus of the cell with the chromosomes, but is found in the mitochondria, which are little oval organs that float around in each cell and have the job of converting nutrients into energy. What is special about mitochondrial DNA is that it comes only from the mother.

Geneticists are fascinated by mitochondrial DNA because they believe that we can use it to get a genetic picture of the "first mother" for any species. In theory, mitochondrial DNA can be used to trace a simple line of descent from female-to-female-to-female back to the beginning. This DNA, however, does slowly change over generations. Over time, mutations occur due to copying mistakes or damage from natural radiation and exposure to chemicals. This means that as time passes the DNA of generations of individuals will drift away from being an exact copy of the original mother's genetic material. Suppose that at some point in time evolution allowed a single entity to separate into two species, races, or breeds. As many years go by, the mitochondrial DNA of the two diverging lines is expected to become more and more different. The original ancestors can still be clearly identified from the mitochondrial DNA, because clusters of mutations are not shuffled into new combinations, as are the genes on chromosomes, but remain together in a particular sequence. In effect, the mutations found in each species become a signature of that line of descent. The degree of difference between the DNA of two species will indicate how long it has been since they went their separate evolutionary ways.

When this kind of genetic analysis was first used, some scientists suggested that the first domestication of wolves that resulted in the creation of dogs may have taken place more than 100,000 years ago. This didn't make sense to many researchers, since it appeared to be far too early. The humanoid species that would have had to do the domestication would have had to have been the Neanderthals, but the fossil evidence makes it clear that Neanderthals never lived in cooperative association with dogs and never seemed to have domesticated anything. Newer research, however, dates the first domestication of wolves or wild dogs at around the end of the Pleistocene era, or around 15,000 years ago, which would be more consistent with actual physical findings of ancient dog bones.

What appears to be the earliest unambiguous fossil evidence of domesticated dogs was uncovered from the Bryansk Region in the central Russian Plain, roughly 400 miles southeast of Moscow. Based upon radio-carbon dating, these two skulls from domesticated dogs appear to be at least 13,000 years old, and may be as old as 17,000 years. Reconstructions based on these remains indicate that these "first dogs" looked much like modern Siberian Huskies, only with a broader, heavier head and muzzle.

The mitochondrial DNA studies also confirm that the "first mother" of dogs was probably a gray wolf, and this wolf's DNA is the closest to dogs'. However, "first" in this instance doesn't mean "only." The scientists who concluded that dogs were domesticated from wolves suggest that the great variety of sizes and shapes that different dog breeds display comes from the fact that at various times different local strains of wolf were domesticated. This is supported by the DNA evidence which suggests that the domestication of wolves occurred in at least five different places at different times, starting in Asia and later in Europe. There were also at least three different times and places in the Americas when the wolf was domesticated.

The DNA evidence makes it clear that there is more to dogs than wolves. While the wolf was the wild canine species most often domesticated, jackals, wild dogs, and coyotes were also likely tamed at later dates and their genes thus enter into the mix that makes up our dogs. In other words, wolves may have been the first canines to enter our homes and share our food and fire, but other members of the canid family that happened to be around also became candidates for domestication. It is certainly the case that domestic dogs can successfully mate with wolves, coyotes, dingoes, jackals, wild dogs, and even some types of foxes (although not the common red fox, which has the wrong number of chromosomes). Some biologists contend that this ability to mate and produce live fertile offspring is proof-positive that individuals are of the same species, which would make our domestic dogs a mix of all of those other wild canines. So the dog resting near you might actually be a "wolf-jackal-coyote-dingo cross." Talk about mixed heritage.

Domesticating Dogs: Why Bother?

Even if the DNA evidence tells us what and where dogs came from and approximately when they were first domesticated, it does not tell us how domestication was accomplished. This process had to involve modifying the personality of wolves and other wild canines. Remember that personality is the key to predicting behavior. The speculation goes something like this. Although the wild ancestors of dogs were efficient and daring pack hunters, they were also opportunistic scavengers. Scavenging is a lot less dangerous and effortful than hunting, especially when the creature is hunting larger animals. So when the opportunity to snatch a free meal from the leavings of another hunter presented itself, these early wild canines were quick to take advantage of it.

Fortunately for the development of future relationships between people and dogs, early humans generated lots of garbage from unused scraps and bones coming from the animals that they killed. Such leftovers were often dumped just outside the camp or settlement in heaps that archeologists call "middens." This free food kept the ancestors of dogs near the camp. Humans tolerated the canines initially, simply because the canids consumed most of their garbage, which reduced the smell and the number of vermin, making the area more pleasant to live in.

Later humans found that the canines came to view the area around the camp as their territory, so that when a threatening wild animal or a marauding band of strangers came close to the encampment, they would set up a commotion. The noise gave enough warning for the inhabitants of the camp to rally some form of defense. Thus in those hostile and dangerous times, canids served as ever-vigilant sentries, especially during the dark hours of the night. In this way, the simple presence of canines made life safer for the human residents.

Once the usefulness of this warning function was demonstrated it is likely that some primitive man realized that if canines around the village would sound the alarm at the approach of potential danger, one actually in his family's hut would provide similar warning when anyone or anything approached. This could provide extra protection at the personal level for him and for his family, which might have been what motivated some human to "adopt" a wolf puppy, take it into his tent or hut and tame it.

For primitive humans moving northward in the temperate zones, dogs provided an unexpected survival advantage. The only garments that humans wore in the Stone Age were made from animal skins. Unfortunately, as these hides dried they became stiff and hard. To make them soft enough to wear as clothing and to provide warmth by clinging close to the body required processing. Since the tanning process to produce leathers had not yet been developed, the skins were typically processed by women chewing the hides to work in enzymes that would keep them supple. Later humans learned to rub fat into the skins and laboriously pound the hide until the water was replaced with the fatty oils. Both of these methods involved many days of effort to produce the materials for a single, not particularly warm, garment.

Dogs, however, provided the potential for a biological heat source. To begin with, their body temperature is higher than that of humans (101 degrees Fahrenheit, or 38.3 degrees Celsius). In addition, they like to crowd together, or huddle close to people, when they sleep. Thus, a few dogs could warm a small shelter when kept inside, and serve as a sort of heating blanket at night. Anthropologist Richard Gould studied the Australian aborigines in the late 1960s, and found that they still used tamed dingoes as heat sources in this way. Some arctic tribes in Russia, such as the Samoyeds (who gave their name to the handsome white Nordic dog) commonly do the same. Early in the twentieth century, in Newfoundland, the practice was to tuck a dog under the blankets to warm the bed on cold nights, and thus came the tradition of describing a cold period as "a three-dog night," meaning that it was so cold that you would need three dogs to keep warm.

Humans may have also brought dogs into the home because puppies simply look "cute." Animals and humans alike instinctively feel a special fondness for the young of their own species. Naturalists, such as the Nobel Prize winner Konrad Lorenz, have suggested that this feeling may be triggered by something about the appearance of young animals. In essence, babies and puppies are "cute" because they are small and have big eyes, round flat faces, appealing facial expressions, and make high-pitched sounds. Evolution takes advantage of this "cuteness" since it has survival value. Cuteness in the young seems to make adults more protective and solicitous of them. Contemporary psychologists have shown that this cuteness factor crosses species boundaries as well. We tend to feel more warmly toward kittens than adult cats, and chicks appear more attractive than adult chickens. The same goes for puppies as compared to adult dogs, wolves, or other canid species. It is difficult not to want to take home and adopt virtually every puppy that you meet. Early man and, perhaps to a greater degree, early woman with her maternal instincts, would probably have thought that the more puppylike of the canines scavenging the village were just appealing little animals that needed care and affection. This might have prompted them to rescue a particularly cute orphaned or abandoned puppy to take home and begin the taming process.

Whether for reasons of future protection, warmth, or just the affection feelings aroused by their cute, young appearances, young puppies of different canine species were adopted into the homes of early man. From this point on, the personality of these wild offspring would determine their future fate.

Forever a Puppy

It is now becoming obvious to most psychologists that domestication lies at the end of a scale of changing relationships between humans and animals. It begins with *habituation,* which refers to the fact that the wild animal now comes to accept and not fear the close presence of humans. The next step is *taming,* which is when the animal comes to allow a human to look after and to control it safely, at least to some degree. There are many examples of animals being tamed but not domesticated. For example, in some parts of the world, wild elephants are captured and trained to work in logging and transportation. Wild parrots and some species of monkeys are captured and tamed for amusement. The Australian aboriginals tame dingoes that then function much like dogs when hunting and for other purposes. The case of tame dingoes is a good example of the difference between taming and domestication. Adults can usually be tamed, but it is much easier to tame young animals. For this reason, captive breeding is often practiced, in which the offspring of tamed or captive animals are themselves tamed. The important difference between this and domestication is that the breeding is not selective or deliberate. Thus, the tamed dingoes of the aboriginals may mate and their pups may in turn be tamed, but no effort is made to select which animals mate with each other.

In its simplest form, domestication requires selective breeding that involves removing animals with unwanted characteristics from the breeding cycle. At the next level, it involves deliberately promoting the mating of animals with desirable characteristics. Only when selective breeding has been used to modify an animal into something that is better adapted to living with humans and to human use, do we have a domesticated animal.

It appears very likely that humans did selectively breed the animals that would become dogs. The best guess is that they did so based on one particular cluster of personality characteristics, namely friendliness and lack of aggression toward people. Obviously, any dog that grew to be aggressive toward any family member would not be tolerated. Most likely, its first severe bite meant instant execution. Only those dogs that were friendliest and most easily tamed would be kept, given food, protected, and ultimately have the chance to breed. Their puppies would be dealt with in the same way, with the friendliest kept and nurtured and the more aggressive simply killed or driven away. Over generations the nature of these wild canines would change: they would become friendly, safe, and ultimately domesticated.

Notice that in this scenario the survival of the first dogs in the company of humans would depend most upon their temperament and personality. The argument is that all other characteristics that make a dog a dog rather than a wolf simply came about as a byproduct of this. This selection process was sufficient to create dogs in the first place and was used for thousands of years, until humans learned enough about applied genetics to begin to create breeds of dogs with special working characteristics.

To predict a dog's behavior then, we must understand, not so much what a dog and wolf have in common, but rather how a dog differs from a wolf. The technical term that describes dogs in relationship to wolves is *neoteny*, which refers to the fact that certain features normally found only in infants and young juveniles persist into adulthood. In essence, our domestic dogs are the Peter Pans of the canine world. They are perpetual puppies. Many physical features of an adult domestic dog resemble those of a puppy more than an adult wolf. The most obvious physical differences are the dog's shorter muzzle, wider and more rounded head, somewhat smaller teeth, and floppy ears (which are seen in wolf pups but never adults). However, the most important aspects of neoteny are behavioral.

In their behavior, dogs grow up showing more puppylike behaviors than you would find in a wild canine adult. One demonstration of this is the dog's life-long desire for play. Another is the fact that dogs bark. Barking is a behavior that humans find useful, since dogs are expected to keep watch and to sound the alarm at the approach of strangers or the occurrence of something out of the ordinary. In contrast, adult wolves rarely bark. Wolf puppies do bark (and also whine and whimper), and this set of characteristics is what is carried over into adulthood in our domestic dogs.

Adult wolves are also *xenophobic,* which means that they are fearful of strangers, whether those strangers are human or canine. Wolf puppies, however, are more trusting and will approach strangers confidently and in a friendly manner, although they lose this characteristic fairly quickly. After only four or five weeks of age, the wolf pup becomes more wary and cautious. Friendliness and approachability are characteristics that are prized in domestic dogs kept as pets, and a dog that does not show these puppylike characteristics is not apt to find acceptance in most human homes.

Still another important puppy feature is acceptance of leadership. Under normal circumstances dogs do not actively compete for higher status, and they take direction from individuals who appear to be in control. Like puppies, they seldom challenge the leader or act aggressively toward him. Without these characteristics it would be extremely difficult to train dogs to respond to human commands.

All of these puppy characteristics are valuable for a dog whose ecological niche has been shifted away from living in the wild to living in the more or less constructed and crowded world of people. While these characteristics are lost in wolves as they mature, they persist in dogs throughout their lives.

Secrets from a Silver Fox

The Russian geneticist Dmitry K. Belyaev observed that domestication and neoteny seem to go hand in hand, not just in dogs, but also in other species. The reduced levels of fearfulness and aggression are vital for any domestic animal if it is going to be controlled by humans and taught to work with them and interact with them. Belyaev wondered if, during the early process of domestication, the selection of animals for breeding and taming might have been based upon exactly these behavioral or personality characteristics. As the head of the research group at the Institute of Cytology and Genetics of the Siberian Department of the Russian Academy of Sciences, in Novosibirsk, he and his team of researchers tried to turn back the clock to the point where active efforts to domesticate dogs began. Belyaev reasoned that he could use a wild canine species to "replay the process" and carefully study what was happening during the creation of a domestic dog. At the time of this writing, this project has been in progress for close to fifty years, having begun in 1959. Belyaev lived long enough to see results suggesting that his general notion was correct, and Lyudmila N. Trut has continued the research in the years after his death.

Actual experimental studies on the process of evolution are difficult to set up and conduct. For example, Belyaev's ideal would have been to start with gray wolves, since we believe that they were one of the first domesticated canines. However, when it came to the selection of which wild canine to use as his "protodog," Belyaev decided not to use wolves because wild wolf stocks are no longer genetically "clean." This is because domestic dogs have often escaped and bred with wild wolves, which would make any genetic analysis of the domestication process less clear. Instead, Belyaev chose a canine species that is very close to dogs but cannot naturally breed with them, namely the Russian silver fox (*Vulpes vulpes*). These foxes had been successfully raised in captivity, and tamed to some degree, but they had never before been domesticated. Using this species also provided an important nonscientific benefit. Silver fox fur is long, dense, and soft, and has a subtle silvery shimmer that highlights the dark or medium gray that is the dominant color. For this reason it is prized for making coats and other pieces of fur apparel. Russian science (at least those aspects of science not oriented toward more technological products) has traditionally been poorly funded. Because domesticated foxes would be easier to raise in fox farms, Belyaev obtained some applied agricultural funding. Excess animals not needed for the further research could be sold for their pelts or as breeders, which would also help further fund the research.

Belyaev's actual experiment is conceptually very simple, but involves a lot of work and patience. Beginning with 130 undomesticated foxes, Belyaev established a systematic breeding program. Each new litter of foxes was tested for friendliness toward humans. The initial group of foxes, as you would expect, were difficult to handle, very afraid of people, and generally behaved like wild animals. Testing was quite simple. An experimenter would offer food to each fox kit at the age of one month. At the same time, he would try to touch, pet, and handle it. The test was repeated twice, while the kit was alone and while it was with other fox kits. Each month the testing was repeated until the fox was seven to eight months old. Then each fox was assigned to one of three classes based on how friendly and accepting it was. Foxes that attempted to escape or hide from experimenters, as well as those that tried to bite them, were assigned to Class III. Foxes that were not overtly friendly to the experimenters, but at least allowed themselves to be touched and hand fed, were assigned to Class II. Foxes that were friendly toward the experimenters, and would approach them and allow physical contact, were assigned to Class I. The total number of foxes in Class I and II was only around five percent in the earliest generations. Because Belyaev's hypothesis was that domestication came about because of selective breeding for tameness, friendliness, and sociability, only those foxes in Class I and II were used for breeding. As the numbers grew larger, soon all breeding was done with foxes in Class I.

After only six generations of breeding for friendliness, the researchers added a new grouping, Class IE, which they called the "domesticated elite." The foxes in this class showed very doglike behaviors. They would actively seek out human contact by approaching with their tail wagging and would seek human attention by whining. In each successive generation, only the most tame and friendly foxes were kept. It was apparent that successive generations were becoming more and more like domestic dogs in their behavior. By the twentieth generation, thirty-five percent of the foxes were classified as domesticated elite, and today, after more than forty generations nearly eighty percent of the foxes fall into this category and show many doglike behaviors.

When Is a Fox a Dog?

Clearly, you can genetically manipulate animals to be less fearful, less aggressive, and more sociable by selectively breeding for friendliness. However, a friendly wolf or fox is not necessarily a dog. Dogs often look physically different from wolves and foxes; they have more puppylike behaviors and they develop and mature more slowly than wild wolves and foxes. Does breeding for friendliness bring about any of these changes?

Belyaev expected to find physical as well as behavioral changes in his domesticated fox-dogs. He reasoned that manipulating aspects of behavior that are under genetic control also alters the internal chemistry of the animal changing the concentrations of neurotransmitters and hormones. For example, the breeding experiment showed that over the generations there occurred a steady drop in the hormone-producing activity of the domestic foxes' adrenal glands. The adrenal glands produce corticosteroids, the chemicals associated with the so-called fight-or-flight behaviors (aggressive or fearful responses). It should not be too much of a surprise to find that selectively breeding animals that show less fearfulness and aggression reduces the production of those hormones that affect these behaviors. The genes that control the production of these chemicals also control when specific chemical changes appear as the animal matures. This, in turn, alters the timing of various stages of growth. Such modifications in internal chemistry can also change the very shape, size, and appearance of the body.

Domesticating the foxes also caused an increase in the levels of serotonin in their brains. This result is rather exciting to researchers, since serotonin is associated with mood and emotions. Many antidepressant drugs, such as Prozac, work at increasing levels of serotonin in order to elevate moods. Higher serotonin levels are associated with fewer mood swings and lower aggression; lower levels of serotonin are associated with slower learning rates and less efficient memory. These findings suggest that the process of domestication may actually be producing animals that are effectively more intelligent since they learn how to cope with their environment more quickly and better remember what they have learned.

In addition to changes in behavior and neurochemistry, Belyaev's domesticated foxes also showed physical changes consistent with neoteny, becoming more puppylike in appearance. After several generations of breeding for friendliness, researchers found that an increasing number of the tame foxes had floppy ears and curly tails. Although these characteristics are common in many dog breeds, they are only found in young wolf and fox pups, not adults. Ears normally straighten into their usual pointed shape as the pup grows older. Adult wolves and foxes typically have straight tails that are carried at a downward-pointing angle, as opposed to pups and later generations of fox-dogs, who, like many adult domestic dogs, have tails that are carried high and may be curled over their backs. In later generations of selectively bred foxes, the shape of their skulls began to change, and their faces became more puppylike with shorter, wider muzzles. They also showed increased levels of puppylike behaviors, such as barking. Other new characteristics seen only rarely in wild foxes but common in domestic dogs began to show up. Unusual coat colors began to appear after only ten generations, with some of the domesticated foxes showing piebald or brown mottled coats, or large patches of black and white.

None of these physical characteristics was selected for deliberately, but appeared as a consequence of simply breeding selectively for friendliness. Belyaev suggested that many of these changes are due to the fact that this form of selective breeding slows the rate at which the animals mature. If that happens, then even when the body chemistry finally does change to a more adult pattern it is too late for it to have much of a physical or neurological influence. In effect, domesticated animals become stuck or stalled at some puppylike stage of development.

Belyaev's fox farm experiment proved that selective breeding to create a sociable animal that is neither overly fearful nor overly aggressive is enough to change a wild canine into something like a domestic dog. It also showed that changes in personality and body shape could be brought about rather rapidly, over decades rather than over centuries. The changes in the hormonal and neurochemical systems that come about when breeding for friendly behaviors can delay, extend, or eliminate certain stages of physical and mental development so that the domestic animal never comes to have all of the characteristics normally found in wild members of his species. The experiment's results confirm the theory that, as we humans domesticated our dogs, we also stalled the development of certain aspects of their physiology and their personality so that they remain perpetually in a puppylike, or juvenile, stage.

Are these domestic elite foxes now the equivalent of dogs? By now more than thirty-five generations have been bred, involving some 45,000 foxes. Eventually the scientific team found that they had a surplus of "domesticated foxes," with many more than were needed for research. At the same time, they still faced decreasing research funding due to a weak Russian economy. Getting supplemental research funding from the sale of fox pelts was difficult since the changes in coat colors had made them less desirable to furriers. A partial solution to the problem was to sell the surplus animals (with the unfashionable pelt colors) as pets to provide some additional funds to keep the research project going. While this was an economically driven action, it did offer the scientists a chance to see how closely they had come to duplicating domestic dogs since they could monitor a number of their domesticated foxes to see how they were doing when adopted into human homes.

It is unlikely that a casual observer would recognize Belyaev's animals as foxes. They appear to be some kind of interesting mixed breed of dog. When they were adopted into typical human households, these domestic foxes did well, not only looking like but also demonstrating personalities and behaviors that are much like dogs. Their owners described them as being "good-tempered" companions and pleasant pets. They seek human company, lick faces, bark, whimper, and wag their tails. They bond well with humans although they are a bit more independent or catlike than most dog breeds.

Are these domesticated foxes now effectively (if not genetically) dogs? That may be more of a philosophical than a psychological or biological question. I think, however, that it is safe to say that they are no longer silver foxes, just as we can say that a Pekingese or Doberman pinscher is no longer a gray wolf.

CHAPTER THREE

Not a Wolf in Sheep's Clothing

Some ideas do not die easily, and the idea that a dog is simply a tamed wolf has been in the scientific and popular literature for so long that it is difficult to erase it. Because wolves and dogs have a shared ancestry, we have a fair understanding of the similarities in their behaviors, but scientists and the general public appear to be less aware of their significant differences. In this chapter, we'll explore not only which characteristics and behaviors are different in wolves and dogs, but also the degrees of those differences.

One of the major differences between dogs and wolves is their personalities. Dogs are typically sociable and friendly, while wolves are more fearful and aggressive. Results from the Siberian fox experiment suggest that these personality characteristics are under genetic control—but to what degree? Can these traits also be modified by a wolf's environment and its experiences—such as the amount of contact that it has had with humans since puppyhood—in the same way as they can be changed by direct genetic manipulation? In other words, can we rear the puppy of a wild wolf in such a way that it has the personality and behaviors of a domestic dog? The simple answer is "no," but there is a good deal of history and research behind that conclusion.

Since domestication occurred so long ago, during prehistoric times, many of our beliefs about people's early relationships with dogs, wolves, and wild dogs are sheer speculations. In some respects, we have not moved very far from the vision of the British writer Rudyard Kipling in 1912 when he offered his theory of the domestication of dogs in his *Just So Stories.* The story begins with the wild dog/wolf/jackal/coyote hanging around the home of the humans, looking at the food being cooked by the primitive human female, and feeling hungry.

"Then the Woman picked up a roasted mutton-bone and threw it to Wild Dog, and said, 'Wild Thing out of the Wild Woods, taste and try.' Wild Dog gnawed the bone, and it was more delicious than anything he had ever tasted, and he said, 'O my Enemy and Wife of my Enemy, give me another.'

"The Woman said, 'Wild Thing out of the Wild Woods, help my Man to hunt through the day and guard this Cave at night, and I will give you as many roast bones as you need.'"

This is still the most common view (minus the talking wild dog, of course) of how wolves became our dogs. The language used by many people studying the origins of dogs, such as "taming" rather than "domesticating" wolves, would be consistent with Kipling's bit of fantasy and suggests that the only differences between the docile pet at your feet and a wild arctic wolf baying at the moon is that your dog has agreed to a living arrangement with humans. In exchange for his good behavior, his wearing a collar, and eating out of a bowl, we will take care of him and support him for life. The implication is that, if he wanted, he could simply head north and join with his wild comrades to hunt elk and howl the night away. The flip side of this coin is the presumption that the wild wolf could head south, put on a collar, and live the life of a Labrador retriever whose only job is to serve as a family's pet, if only he chose to give up the freedom of his wandering ways.

A Wolf in the Parlor

Belyaev's study of silver foxes demonstrates one way that wild canines can be domesticated. If this process roughly reproduces the way domestication was accomplished by primitive humans, domestication may have begun with genetic manipulation. Since early humans knew nothing about genes, this would have been accidental manipulation, as friendly and tame animals are preferred, and then allowed to breed with each other. Dogs would then be the result of many generations of selective breeding that ultimately caused their personality to become acceptable and desirable to humans.

Yet Belyaev's conclusion doesn't agree with the more common view, like Kipling's tale, in which wild canines began hanging around human settlements eating garbage until some early human took one of their pups inside, fed it, and it became tame simply because it was reared in the company of humans. According to that view, it is only later, when the now-tame animals were allowed to breed, that humans could have made those changes permanent through selective breeding and culling out any dogs with aggressive, fearful, or other undesirable characteristics.

We need to conduct research that is the reverse of the fox farm study to decide if simply adopting a wild wolf pup is a likely route to domestication. Belyaev manipulated the genetics of his foxes, but every generation had virtually the same environment and similar rearing experiences. In this alternative line of research, we need to take wild canines that have not been selectively bred and see if they can be tamed using only behavioral methods. A number of researchers have done such research, trying to rear wild-born wolves in captivity, to see how much they could change their personality and behaviors by controlling their environment and interactions with people.

Typical of this kind of research was a study by psychologist Harry Frank at the University of Michigan at Flint, who reared captive wolves with the assistance of his wife Martha. The Franks obtained a number of eleven-day-old wolf pups and tried to raise them in such a way that they would develop a social, affectionate bond with their human caregivers. These researchers spent approximately twelve hours of each day with the wolf pups, and even hand-fed them using a bottle until the age when normally they would be weaned by their mother. Until the pups were around seven weeks of age, they alternated nights sleeping with their mother and sleeping in the Franks's home.

Despite this intensive contact, the wolf pups developed only a tenuous relationship with the researchers who nurtured them. When they were young, and unfamiliar humans approached them, the wolf pups would seek out an adult wolf or one of the researchers' adult dogs to hide behind. As the wolves grew older, they still did not act like dogs and form a social bond with humans. For all of their lives they continued to show a distinct preference for canine social partners rather than their human foster family.

All studies that have attempted to tame wolves have shown that a set of highly stable emotional responses shapes their personality. In the wild, the wolf's personality seems to be dominated by a general suspiciousness, which often leads to fear-driven, self-protective behaviors. Their social behaviors are oriented toward dominance and status in their pack. Aggression is easily triggered in wild wolves and may be brought out by fear, the desire to exert dominance, the desire for a particular item of food, an object, or even conflict over a particular place to sleep. Living beings that are not threats, or are obviously weak or injured, will often bring out the predatory instincts in wolves. This will show up in the form of stalking, threatening, or outright attack of the weak or injured. These kinds of personality traits are incompatible with taming wolves and establishing a doglike bond with humans.

Some Things Are Hard to Change

Research confirming the difficulties in modifying a wolf's personality traits was conducted by John Fentress, who is associated with Dalhousie University and the Canadian Centre for Wolf Research. Fentress wanted to see if a wolf pup that was raised much like a pet dog from the age of four weeks would grow up to have doglike personality traits. To test this idea he took a wolf pup, named it Lupey, and reared it in his own home, exposing it to the same kind of human family environment that domestic dog pups experience. At the scientific level, the object is to see if learning and experience alone could change a wolf's genetically determined personality.

The first thing that Fentress noticed was that the wolf pup didn't seem to be as flexible or adaptable as a typical dog pup. During his first few weeks, Lupey had difficulty adjusting to a new diet and had to be hand-fed until he eventually got used to his new food. Over time, the pup became sociable with family members and other familiar humans but remained very cautious and suspicious of strangers. Surprisingly, as he grew older, his wolfish personality, instead of softening, became increasingly more apparent. By the age of thirteen weeks, the pup began to show predatory aggression and killed chickens if given an opportunity. Furthermore, when Fentress tried to remove the dead chicken, he was met by direct aggressive threats. At fourteen weeks, Lupey was regularly killing rodents (one of the mainstays of wild wolves' diet). When brought near larger animals, particularly horses, he would become extremely excited and try to nip at their tails.

When Lupey was six months old, his interactions with dogs were friendly, and he would invite them to play without showing any overt aggression. However, at this age he also killed a family cat that he had been raised with and which had never acted aggressively toward him. Nonetheless, Fentress was not overly concerned since Lupey did not seem to be a danger to anything else and was still friendly with familiar humans.

By the time the pup had reached one year of age, however, he began to act even more like a wild wolf, becoming more aggressive in his hunting behavior and routinely attacking cats, chickens, and geese if the opportunity presented itself. Earlier in his life, he had been given some basic obedience training, much like a family dog, and had learned a few commands. By the age of one year, however, he was much less likely to perform these learned behaviors on cue than when he was younger. He also seemed to become more independent and restless and started to perform what looked like hunting maneuvers, often involving mock attacks and pounces—occasionally with humans as the target. Fentress did not consider these behaviors actual threats since Lupey still seemed sociable and friendly with family members and humans to whom he was accustomed.

By the time the wolf reached his second year of age, his attitude toward small children had changed. Now the wolf began to watch the children and stalk them with the same intensity as he normally stalked cats. At this point Fentress did become concerned. He interpreted these behaviors as clear indications of a likely danger and threat and made sure that Lupey was no longer given unsupervised access to children.

At the age of three years, when the study drew to a close, Fentress observed that the wolf remained sociable toward familiar adult humans and dogs with whom he had been reared. The bond with people was weak, however, and if given the opportunity to run freely in a fenced field, he would spend considerable time actively avoiding direct contact with humans. Lupey also remained generally anxious and suspicious when exposed to unfamiliar people or situations, and he continued to be easily frightened and difficult to calm. Fentress ultimately concluded that a wolf can be tamed to be sociable with familiar adults but remains much more fearful and aggressive than a dog and is thus too unreliable to be a pet or companion.

The experience that Fentress had with Lupey is not unusual. Jerome Woolpy of the psychology department of the University of Chicago, observed a similar pattern when his team tried to tame wolves. Early in the process fearfulness seemed to dominate the wolf's behaviors, although, as the wolf became more familiar and confident around the researcher his behavior began to change, but not in the desired direction. Generally, as a wolf's confidence and familiarity with the humans in his environment grew, so did the likelihood that he would use aggression to control the behaviors of those around him. As the wolf became bolder, he began to bite and tug at any protective clothing his caretakers were wearing. If researchers tried to prevent his biting at their clothes, he would bite harder and more vigorously. If the researchers went further and attempted to dominate the wolf physically at this stage, it would frequently lead to full-blown attack. If the attempt to dominate the wolf succeeded, it seemed to set the wolf back to an earlier stage in the taming process where fearful responses and suspicion again seem to be the most common behaviors. If, on the other hand, the researcher retreated too quickly after an attack, the wolf seemed to grow more courageous and inclined to show even greater levels of aggressive behavior toward humans in his next encounters with them.

Even in cases in which wolves were bred in captivity and their taming and socialization process had begun virtually from birth, the bond with humans was easily lost. A few months with no active interactions with people resulted in animals that acted as if they had never been tamed, and their behavior again seemed dominated by fearfulness and easily triggered aggression.

Love Between Wolves and Poodles

Erik Zimen chose a different method to explore differences between the personalities of wolves and dogs. Initially interested in taming wild wolves in order to understand how the process of domesticating dogs might have come about, he set up his first research facility in a former forestry station in northern Germany. However, when he encountered the same difficulties that earlier researchers had, he began to think about genetically determined behaviors that can't be changed by rearing practices. This led him to study what happens when dogs and wolves are interbred. The first wolf that he "adopted" was a female who was taken from her litter at the age of twenty-one days and hand-raised. Like Fentress, he tried to tame her by rearing her in his home, but the outcome was just as unsuccessful. When she was young, the wolf showed evidence of being closely attached to him. However, as she matured and, some time after she reached sexual maturity, she turned against him so violently that he had to break off contact with her. Zimen's later research led him to conclude that the emergence of social aggression as the wolf grew older was the most common pattern of behavior development seen when trying to tame wild wolves of both sexes. To his surprise, the data also suggested that this tendency was stronger in females.

In later studies, Zimen found that the only way to have a chance of successfully socializing wolves to humans was to start rearing them in a human environment, in the absence of other wolves or dogs, from the age of fourteen days. This is extremely young and is just about the age that wolf pups begin to open their eyes. Obviously, such pups would be too young to be fed solid food so they had to be bottle-fed. In addition, for the socialization process to have a strong hold on the wolves' personality, they had to be in the presence of their human caregivers virtually all of their waking hours for the first few months of their lives. All of this suggested to Zimen that the genetic influence on personality was extremely strong and generally not compatible with easily taming a wild canine to make it a suitable companion for a human family.

To investigate this issue further, Zimen began some genetic manipulations. Specifically he cross bred wolves and poodles. Anticipating the fancy names that are currently being used for the deliberately crossbred dog breeds that we have today (such as *labradoodles,* for Labrador retriever and poodle crosses, and *cockapoos,* for cocker spaniel and poodle crosses), he called the offspring of his poodle-wolf crosses *puwos.* Soon after birth each litter (whether wolf, dog, or crossbred), the puppies were taken from their mothers and placed together in separate enclosures to prevent them from acquiring any learned behaviors from their mothers.

The wolf-dog hybrids demonstrated that the genes that determine fearful and aggressive personalities in wolves seem to be dominant over the friendly and sociable genes in domestic dogs. The first generation of wolf-poodle hybrids, even though hand raised, had personalities and behavior reactions that were very similar to purebred wolves. There were frequent conflicts among the puppies in each litter, which seemed to involve attempts at establishing social dominance. In general, Zimen observed that the dominance battles among the females were fiercest and more likely to lead to actual physical damage. There were differences, however, between the dog litters and the wolves or wolf-cross litters. The dog puppies established a dominance order among themselves that remained fairly fixed. The dominance relationships among wolf pups seemed more complex and changing. The wolf pups and the hybrids constantly squabbled and tested their littermates for strength and control.

Zimen observed that, like wolves, before they reached twenty-one days of age, all of the hybrid pups began to show fearfulness when approached by humans and reacted by trying to get as far away from them as possible. Eventually, these hybrids did become socialized to the point where they did not act fearfully around Zimen, but they never became socialized toward strangers and the appearance of any unfamiliar human usually provoked attempts to escape or hide.

The first generation of puwos contained a mix of dog and wolf characteristics with a bias toward wolflike personalities. When they were allowed to interbreed among themselves, the second generation showed less uniformity. Their behavior patterns randomly differed among the members of the same litter. Different pups displayed varying degrees of emotional responses to an approaching human. These ranged from the extreme fearfulness characteristic of wolves to the friendly approach of domestic dog puppies. These early response patterns were good predictors of the ultimate success of attempts to socialize the pups. Those that showed wolflike fearfulness never became comfortable around humans, but those that showed less fear, and especially those that showed interest in social contact with people, could be socialized, eventually becoming friendly and excited when humans approached.

The fact that the temperament that the pups display when still in their litter is such a strong indicator of whether, as adults, they will get along comfortably with humans suggests that personality has a strong genetic determination. However, early experience can play a role. When puppies that were originally friendly and less timid were allowed to remain with a mother that was anxious around people or with hybrid littermates that were extremely fearful at the approach of humans, their behaviors changed. By the age of six weeks, these originally sociable pups showed the fearfulness and escape tendencies typical of wolves. Thus, the more desirable personality characteristics that we find in domestic dogs (friendliness, confidence, nonaggression) can be lost because of early experiences. Zimen also noticed that, even when puppies were raised by a friendly and well-socialized mother, those hybrids that were most wolflike in their behaviors still remained nonsocial and shy around humans. Thus, doglike personality traits can easily revert back to the temperament traits found in wild canines while the wolf's fear and aggression seem firmly set from birth.

Dangerous Pets

The kind of research that Zimen conducted has clear implications for the increasing numbers of people who accept so-called tame wolves or wolf and dog crossbreeds as pets. This pet wolf fad is quite widespread, and suppliers of wolves and wolf-dog crossbreeds can easily be found on the internet. There may be between 300,000 and 500,000 of such wolf-dog mixes in the United States. In general, these hybrids seem to appeal to many individuals who want an exotic breed of dog as a sort of trophy or to make some sort of macho statement about how powerful, threatening, and dominant they are. However, there are some people who claim that they want such wolf hybrids because they have desirable characteristics that domestication has taken out of our domestic *Canis familiaris* while producing breeds like the Labrador retriever or golden retriever. Breeders and owners of wolf-dog crossbreeds argue that these animals are more intelligent, independent, complex, and resourceful than their fully domesticated cousins, yet have all of the characteristics to become a good family dog despite the scientific findings which suggest that this is not the case.

Even if we did not have Zimen's data, which suggest that in wolf-dog crossbreeds the wolf traits of aggressiveness and fearfulness are the most likely to dominate hybrids' personalities, I would still be wary of them. Such a wolf-dog cross is bound to undo or at least weaken the effects of 15,000 years of breeding for friendliness and sociability. Certainly, as pups these wolf hybrids behave playfully, like dogs; they may be a bit rough, but their play is mostly friendly. However, the data indicate that hybrids develop much like the purebred wolf pups that researchers unsuccessfully tried to tame. As they approach physical and sexual maturity—at around two years of age—the more wolf-like aspects of their personality begin to emerge and they begin to act like the highly territorial predators that their wild wolf genes predispose them to be.

Most wolf-dog hybrid owners agree that their animals treat humans as if they are other wolves and demonstrate the genetically programmed predisposition to struggle continuously with pack mates for food and leadership. Furthermore, they often have no inhibitions about challenging the dominant householder by using warning growls and even a succession of bites when the issue is a piece of roast beef on the human's plate or even attention from his spouse. If they are physically punished they will retreat, but like the wild wolves whose genes they carry, they will renew their challenging behavior an hour, day, or week later in what appears to be an unending contest for dominance.

Lorna Williams, who lives in a rural suburb just outside of Chicago, told me of her experience with a wolf-dog hybrid named Bleiz.

"Bleiz was really good around the family, and although Steve [her husband] *and the boys* [her two teenaged sons] *sometimes had to disci-*

pline her for being pushy or uncooperative or grabbing at food and such, we never worried about her. She had just turned about four years of age, when it all fell apart. Steve came into the yard where the boys were helping me with the garden and Bleiz was just sniffing around. Steve had been out jogging and had tripped on something and was limping pretty badly. Bleiz looked at him and without any warning leapt at him, knocked him down, and started to bite him. Fortunately, the boys were out there with me. They grabbed her collar, dragged her away from Steve, and locked her in her kennel run. Steve had protected his face and throat with his arms which were all bloody from dozens of places where the skin had been pierced.

"After that Steve could never approach Bleiz again. When she would see him, she would growl, snap, and try to attack. Our veterinarian put us in touch with an animal behaviorist, who didn't even bother coming down to examine her. As soon as we told him that she was a wolf cross, he said to me, 'When she saw your husband limping, she took this as a weakness and the perfect time to challenge him for leadership. Now that she has found that she can hurt him and put him on the ground, she will never stop her challenges. Your only option is to get rid of her.' Unfortunately, that was what we had to do."

Randall Lockwood, a vice president of the Humane Society of the United States summarized the consensus opinion that most experts have about wolf-dog hybrids when he said, "I would estimate perhaps 80 to 90 percent of owners are experiencing serious problems with their hybrids by the time the animals are three years of age." These problems are almost always aggression and dominance-related and they can lead to fatalities.

Striking examples of wolf-dog hybrids attacking people led to stories of devil dogs, werewolves, and the dark and fantastic eighteenth-century legend of the Beast of Gevaudan. Tales of vicious wolves were very common in Europe, especially between the years of 1600 and 1800, before wolf kills and wolf control programs decimated the European wolf population. One story that has been told many times, and has been frequently embellished, deals with the wolves of Gevaudan. Fortunately, an objective history of the incident was recorded by Father Francois Fabre, the parish priest of Gevaudan. In the three years spanning 1764 to 1767, two large wolves were blamed for killing a number of villagers near Gevaudan, which is in central France. The total count was sixty-four killings, mostly of children who were by themselves, playing or working in fields or gardens unattended. The adult victims were predominantly women, and the only adult males killed were old, frail, or partly incapacitated.

These wolves were abnormally large and had unusually colored coats. One had a large, white throat patch and white on his paws; while the other was reddish colored with some white markings. Neither of these color combinations is normally found in true wolves. The first of the wolves was finally eventually killed in 1766, and was found to weigh over 130 pounds (60 kilograms), much more than the typical 90 pounds (40 kilograms) of the male gray wolves in that region. The other animal that was killed the following year was a female who weighed 109 pounds (49 kilograms) almost 40 pounds (18 kilograms) more than typical female gray wolves.

What makes this case particularly relevant is that recently these killings were investigated by C. H. D. Clarke, former head of the Wildlife Branch of the Ontario Ministry of Natural Resources. After examining all of the available evidence, he concluded that the Gevaudan beasts were in all likelihood not wolves, but wolf-dog crosses. Clarke maintained that skull measurements from the two animals confirmed his wolf-dog hybrid theory and could also account for the unusual coat colorations. Their large size came from the dogs with which they had interbred. In France at that time, it was common to keep huge mastiffs as guard dogs. It appears that a wolf mated with one of these mastiffs and the two wolf-dog hybrids (probably from the same litter) were then responsible for the extended killing spree. Once these wolf-dogs overcame their initial fearfulness of humans, they looked at humans (particularly young and weak humans) as prey.

Is this simply a lurid, tragic, but unique case of two wolf-dogs gone bad? Certainly one would not condemn all Saint Bernards as being vicious and dangerous simply because in real life one of them turned out to be something like Cujo from Stephen King's novel. Unfortunately, the evidence is clear that wolf-dog hybrids are significantly more dangerous than other dogs.

Data analyzed by researchers from the U.S. National Center for Injury Prevention and Control suggest that the wolf's genetically determined personality characteristics make wolf-dog hybrids more dangerous than domestic dogs. Statistics are not readily available for all cases of dog bites, since there is no national registry for this, but all dog-bite-related *deaths* must be reported. Researchers analyzed all such deaths from 1979 through 1998, and found 238 in which the dog's breed was recorded. Of these, 15 were due to fatal maulings by wolf hybrids. To get an idea of what this means, we first have to recognize that during that period of time the dog population in the United States averaged around 56 million, and the best estimate of the number of wolf-dog hybrids at that same time period was about 300,000. This means that wolf hybrids made up just over one-half percent of all of the dogs in the country. If their aggression level is the same as that of domestic dogs, then these wolf crossbreeds should account for one-half percent of all dog bite fatalities (perhaps one or two out of the 238). Instead, the wolf hybrids account for 6.3 percent of all fatalities, which is around 12 times more than the number we would expect based upon their population statistics. This is consistent with the suggestion that wolf-dog crosses are more aggressive and dangerous than the common domestic dog, whose genetic makeup has been modified for many thousands of years to remove much of the wolf's dominance, suspiciousness, and overt aggression.

The wolf research facility, Wolf Park, founded by Erich Klinghammer, has concluded simply, "Wolves and high-content wolf hybrids should never be regarded as pets." Their genetic makeup gives them a personality that is aggressive and predatory. They are unsuitable as companion animals and I for one, would never trust them around children.

Could Og the Caveman Really Tame the Big Bad Wolf?

Considering all that we have learned about the wolf's personality, especially in terms of sociability toward humans, we are left with an interesting problem when considering the usual theory of how dogs were domesticated. The suggestion is that a human adopted the pups of our ancestral dogs that were hovering around their prehistoric camp, tamed them, and then selectively bred them for desirable personality and behavioral characteristics. It was in this way that we started down the road that ultimately led us to the multiplicity of breeds that make our modern domestic dogs.

If we carefully think through this scenario, however, it begins to appear problematic. Wolves' personalities predispose them to avoid humans and run away from them. If their fearful avoidance is prevented, they respond with threats and aggression. A massive amount of time and effort was required to make very young wolf pups receptive to human interaction. In the first stages, the researchers had to sit or stand quietly for long hours so as not to frighten the wolf pups. This had to be repeated each day and for many days on end until the wolves habituated, which means that they stopped showing fear every time a human approached them, or merely moved while standing nearby. This had to be done in some sort of enclosed area to prevent the wolf pups from simply running away. Without a restraint or enclosure, the wolves would continue to spend most of their time trying to escape and avoid their caretakers. Even so, this laborious procedure really results in tolerance of humans, not tameness, and not a truly socialized animal. It seems unlikely that primitive men would have expended that amount of time and effort, especially given the fact that at this stage there was no obvious benefit to be gained by having a tame wolf around.

Some theorists speculating about domestication of dogs have suggested an alternative course of action might have occurred. According to them, wolves were domesticated simply by being fed by humans. This is what Kipling's little fantasy story proposed. The idea is that humans would toss food to the wild canines, and over a period of time they would become less fearful and tamer mostly because they simply became accustomed to the people and grew to trust them. Unfortunately, research shows that without confinement to keep them near humans, free-ranging wolves cannot be socialized. In fact, simply feeding free-ranging animals often leads to very undesirable behaviors. A well-known example of this in another species is the case of bears that obtain food from humans who offer them treats; they soon learn that they can get food from garbage left in parks, campgrounds, or city suburbs. The problem is that once these animals lose their fear of humans, rather than acting tamely, they actually become more dangerous and difficult to control. When they reach this point, they are often willing to use direct aggression, whether to challenge people for desired items, such as food, or simply to express social dominance. The U.S. National Park Service reports that bears that have a history of taking treats and food handouts from park visitors are six times more likely to attack humans than those who have not. There is no reason to expect that wolves would react differently.

Let's look at some other details of the taming theory of dog domestication for a moment. The scientific data suggest that to have any reasonable chance to tame a wolf cub successfully it would have to be captured and taken from its litter at approximately two weeks of age. This requirement alone would make it difficult for primitive humans to rear them, since at this age it would be necessary to supply the pups with nourishment in the form of milk for several weeks. Without bottles with nipples, the only possible way to do this would be to find a lactating human female willing to nurse the pups. Although this appears unusual, it is not completely unheard of. There have been occasional reports that on some Polynesian islands women villagers will nurse pups (or sometimes pigs). Often the purpose of this is not to tame or domesticate the animals, but rather to keep it alive for a few weeks since young puppies are considered a delicacy and may be served as a special dish on special occasions.

Let's suppose that some early primitive woman decided to rear the pup until it was on its own. The amount of work involved would have been daunting. First, wolf cubs (like dog pups) require almost hourly feedings during the first few weeks of life. Their pattern of feeding is erratic as well, and pups will often fall asleep while continuing to suckle and then wake up periodically to start nursing again. A human wet nurse would have had to carry the pup against her breast constantly. There are other requirements as well. For example, puppies don't maintain their body heat very well so that they would have had to be wrapped in a blanket and carried close to their caretaker's body to keep warm.

A more difficult problem is that young pups need to be stimulated before they urinate and defecate, which their mother does by licking them in the anal and genital regions. She then normally eats the fecal material to keep the puppy and the den clean. A human caretaker would need to rub the puppy in the anal-genital regions periodically in order to trigger elimination and clean up the resulting material. Somewhere around three weeks of age the pups usually start receiving their first solid food. For wolf pups, this is provided by the mother who regurgitates some of her partly digested food for the puppy to lick and eat. To mimic this, the adoptive human mother would have to chew the food and spit it out for the puppy to eat. In the wild, this method of obtaining the bulk of its food can sometimes continue until the pup is nearly two months old or until it gets its permanent teeth.

While all of this is possible, we are looking at a process which would be very demanding, both physically and psychologically, perhaps even more so than caring for a human newborn. It would take a good deal of dedication, and there would have to be a really good reason for such an effort. Perhaps a woman whose own infant had died, and who was looking for some sort of emotional solace might have tried it; however, other members of her tribe or village might consider this a very odd and strange form of behavior. Remember, at this time in history we are looking at wild animal whose ultimate usefulness has not yet been established.

Nevertheless, let us suppose that the woman receives sufficient social support and the puppy survives to an age where it can be relatively self-sufficient. There are now additional problems. The juvenile pup would now have to be restrained and confined in some way to continue the taming process. Remember that the researchers who tried to tame wolves found that juvenile wolves that had been reared with humans still tried to avoid the approach of humans. Tethering the animal would be an unlikely solution, since the strong jaws of a wolf would be able to chew through rope or leather eventually. Furthermore, tethering dogs has been shown to increase their aggressiveness, which is exactly the opposite of what is needed if a tame animal is desired. Instead, a sturdy cage or enclosure would be required. Simple fencing would probably not work because the wolf could dig its way out. This seems like a lot of effort and would require more than one distraught mother seeking a child substitute.

Suppose that all of these difficulties were somehow surmounted, and that the tamed animal was a female who survived in the human settlement to maturity. Unless similar effort had also been made to insure that a male wolf cub was also nursed, tamed, and socialized, the problem now becomes how to obtain the next generation of tamed wolves. To begin with, there is a safety issue. As the wolf pup reaches adulthood and becomes sexually mature, the likelihood that it will become aggressive increases. Now the human settlement would have to deal with an animal that is a potential threat to human children and much more likely to challenge adult humans.

Again, let us assume that all of these obstacles are surmounted; there are still reproductive and developmental issues to deal with. Wild wolves only come into heat once a year and most are not reproductively mature until two years of age or older. It might be possible to tether out the female when she was in heat in order to let her mate with a wild male, but wolves are not as promiscuous as dogs. Each pack usually has only one breeding pair. A strange wolf, not a pack member, would more likely be treated as an invader of the pack's territory. Lone male wolves do exist, so suppose that such a male did mate with her. Now the pregnant animal would have to be confined again to have her pups, since, in the wild, pregnant females whelp their litters in secluded and isolated areas. Looking at the best-case scenario, suppose that we did obtain a litter. Remember that the researchers looking at the taming of wolves found that pups born to socialized wolves were just as fearful and suspicious as those born to wild wolves. This means that the labor-intensive process of taming must start again from the beginning with her offspring.

While it is not impossible that such a process resulted in the first domestication of dogs, it seems extremely unlikely. With only one litter born under complete human control it would take many years to select out those with better temperaments and interbreed them to a degree where there was a noticeable change in personality due to genetic influences. Even if the first tame litter was successfully bred, many years and generations would have to pass before a creature that approximates a domestic dog would emerge.

How Dogs Domesticated Dogs

Instead of the traditional "captured wolf cub theory," I would like to suggest an alternate theory for how wolves were domesticated and became dogs. In this theory, much of the selection and domestication comes from the wild canines themselves and not from direct manipulation by humans. In terms of evolution, successful animals are those that survive and increase in population numbers. To do so, animals must be well adapted genetically, physically, and behaviorally to their environment. The environment to which the domestic dog must adapt is one created by and populated with humans.

Let us flash back now to our human settlement where the ancestors of dogs are pawing through the garbage heaps for food. Over time, these particular wolves not only look at this area as their home range, but also ultimately become dependent on this human refuse as their primary food source. Although these animals are tolerated because they remove waste material, any that are aggressive or threaten people would be killed or driven away by the human residents of the settlement. This culling process alone would start to exert some genetic changes by eliminating the most aggressive members of the pack from the pool of available breeders.

Wolves, like the foxes that Belyaev studied, have some individual variations in personality. This means that in the pack of human-refuse-eating canines, some animals would be a bit less fearful and suspicious than others and, living near humans, these would have a definite advantage. Those who are less fearful would not run away and try to hide at the approach of people, but rather might watch warily while continuing to forage for food. This provides the less-fearful canines with two advantages: first, they expend less energy than those who ran, and second, they get more time to feed and select the better, more nutritious morsels. Ultimately, these more sociable animals will be healthier and will be more likely to have offspring. Remember that the personality traits of fearfulness and friendliness seem to be genetically determined to a great degree. So the litters of the more socially oriented wolves will contain a greater number of pups that are more comfortable around their human neighbors and are effectively tamer. Over successive generations, these *settlement-dwelling wolves* will come to prosper and their numbers will grow. Eventually, the tamest of them will be comfortable openly foraging during the day.

The most sociable of the settlement-dwelling wolves would also have gained other advantages. Content in the presence of humans, they would sleep nearer to the village and bear their puppies close by. This affords them, and especially their vulnerable puppies, additional safety since most of the large predators that are a threat to wolves try to avoid concentrations of humans. Wolves that are truly serene around humans might even seek additional benefits, such as huddling against human residences to obtain some of the heat leaking out during the cold months of winter. Over time, these small advantages would add up and increase the survival chances of the most sociable members of the group.

Notice that in effect Belyaev's selective breeding procedure is occurring without much human involvement. The simple geographical separation between the more fearful wolves living in the woods and the settlement-dwelling wolves will make it more likely that friendly and fearless animals will breed with other friendly and fearless animals. Over a number of generations, the originally wild-type wolves will have changed. What I have been calling settlement-dwelling wolves are really animals that have become genetically different from the original wild stock in much the way that later generations of the fox farm animals were different from the original wild silver foxes. The only human intervention might be actions to insure public safety that involve eliminating those now rare individuals that are genetic throwbacks and display a wolfish aggressive personality.

After wolves reached this point genetically, the commonly suggested theory of domestication begins to make sense. The traditional theory can now work because the starting point is no longer with wild wolf pups, but rather with puppies from this new species of settlement wolves, which are partially tamed already. Since the settlement wolves live in such close proximity to humans, when they whelp, their litters will be more likely to be found by humans. Since they are not as fearful and wary as the wild stock, taming them doesn't have to start at such an early age. If the pups can be adopted at an older age, then the burden of early care is considerably reduced. Because their personality no longer has the predisposition to fear and avoid humans, restraint and confinement will not be needed, and these adopted canines can more freely interact with humans. That fact alone will make the socialization process easier. The wild canines that became dogs appear to have started the domestication process themselves. It is from this point on that human interventions can start to further shape the nature of dogs as we selectively mate animals that have desirable characteristics. It is likely that humans recognized what was happening to the settlement-dwelling wolves and took advantage of the partially tamed, partially domesticated animal that appeared as if it might prove to be useful.

The process of domesticating dogs has important implications for understanding the behavior of dogs. It suggests that using the behavior of wolves to explain the behavior of dogs is the wrong method to pursue. Clearly, dogs are not wolves, and even before dogs were domesticated, the settlement-dwelling wolves had evolved to become something between the wolf and the dog. If dogs still had the same personality characteristics as wolves, we would not have domesticated dogs today. This also means that any answer to the question, "Why does my dog act that way?" that replies, "Because wolves act that way," is too simplistic. Dogs and wolves share genes and an early evolutionary history just as humans and chimpanzees share many genes and ancient ancestors. However, humans are not chimpanzees and dogs are not wolves.

CHAPTER FOUR

A Mirror in Fur

I once overheard a snippet of conversation between two women. One was having problems training her little Shetland sheepdog and the other was trying to analyze the situation. The woman with the problem was asked for information about the dog's performance ("She gets frantic and then gives up when I try to get her to learn something new") and then, strangely, she was also asked to give her dog's birthdate and her own. This intrigued me, so I tried to listen in on their conversation a bit longer.

"Ah, that's your problem," the advisor said, as though she had just had an epiphany. "You are a Gemini, and with that astrological sign you are a thinker and act logically. On the other hand, your dog is a Cancer, and with that sign, she is ruled by her feelings, not logic. The two of you have incompatible personalities, and if this were a marriage, I would be counseling you to seek help or consider a separation. The only hope is that, since the dog can't change its point of view, you must get in tune with her personality and start to try to find out what she is feeling when you are training her."

The speakers moved out of earshot so I couldn't hear the specific advice given, but I already knew that I wasn't impressed with her behavior analysis. As a scientist, I really have little faith in astrology (although when I am nonscientifically reading the newspaper I do check my daily horoscope, just like every other scientist probably also does). However, this conversation does illustrate how we tend to think of dogs. It seems as though, at many levels, we can't get away from presuming that dogs differ from humans only in the fact that they move on four legs and sleep in fur coats. We presume that they have personalities like ours, and that these are influenced by the same forces as human personalities (including the stars and planets, if such things are part of our belief structure). We even believe that dogs' personalities will mesh or clash with people, just as any pair of humans might be compatible or incompatible. Before we can determine whether these beliefs are true or false, however, we really have to do some basic analysis of the psychology of personality.

What Is Personality?

As a professor and a scientist, it should seem odd to me that we have been talking about personality without actually defining what we mean by it. We can get away with this, in part, because most people have some notion of what we mean by the term. However, if we are really going to explore personality as it applies to dogs and use it to predict and explain their behaviors, then we are going to have to be more precise.

To begin with, I am not using the term personality in the same sense that it was used in the 1959 rock and roll song by Lloyd Price where he sang "You've got personality," or the way we might use it in casual conversation when we say, "He is very bright, but he doesn't have much personality." In such instances, the word *personality* means a set of desirable characteristics, such as a sense of humor, friendliness, good social skills, some evidence of warmth, caring, and so forth. Psychologists certainly do not mean this when we use the word. If we accepted that definition, someone like Genghis Khan, Attila the Hun, or a vicious serial murderer who lacks any remorse would have to be classified as having no personality at all.

Psychologists use the term "personality" when they are describing individuals based on the presumption that people can be categorized into separate groups according to their typical behaviors. Psychologists refer to this as classifying people into *personality types.* This same label is used in everyday language, such as when we refer to a person as an "aggressive type" or an "ambitious type" or a "loving type" of person.

Sorting people into discrete categories based on their typical behaviors can be traced all the way back to Hippocrates, the Greek physician who lived around 400 BC and who is usually credited with being the "Father of Medicine." Besides authoring the Hippocratic Oath, which in various modified forms is still taken by medical students on graduation from medical school, he also developed what may be the oldest formal theory of personality.

Hippocrates classified people into four personality types. These personality types were supposed to come about because of physiological differences between individuals, specifically the concentrations of four types of bodily fluids, called *humors.* The four humors were blood, black bile, yellow bile, and phlegm, the amount of which determined the individual's activity level, sensitivity, aggressiveness, and overall satisfaction with life. According to Hippocrates, the person with a *sanguine* personality has an excess of blood, which causes him to have a ruddy complexion, but also makes him cheerful and easygoing. The *melancholic* personality has an excess of black bile, which makes him depressed and moody. The *choleric* personality type has an excess of yellow bile, which results in aggressive and excitable behaviors. Finally, the *phlegmatic* personality type has an excess of phlegm, which causes him to be calm and unresponsive.

Of course, no contemporary scientist believes that personality is determined by these four humors, since Hippocrate's primitive interpretations of body chemistry have been superseded by our modern understanding of neurophysiology. Certain aspects of Hippocrates' theory, however, such as its suggestion that body chemistry can influence personality, have survived in some form for more than 2,500 years. The personality categories that he described have proven to be useful enough so that they have become part of our language. To be sanguine about something means to be comfortable, happy, and optimistic about the situation. The term melancholy is used to mean sadness and depression as in the old standard torch song "Melancholy Baby". We still use the word phlegmatic to describe a person who doesn't say much. The slang phrase "hot under the collar" is used to describe someone who is angry and acting aggressively. It is derived from the phrase "hot and under the choler" meaning that a person was acting like a choleric personality would.

The modern personality theorist looking for physical factors that might cause particular behavior patterns is most likely to study hormones, neurotransmitters, and chemicals produced within the brain, such as endorphins, or to discuss the contribution of genetics. Nevertheless, Hippocrate's four personality types are still referred to by some scientists such as Hans J. Eysenck, a German-born British psychologist who is well known as a personality theorist who studies the biological basis of behavior. He reconsidered Hippocrate's four personality types in light of what we know about human behavior today and concluded that division of people into sanguine, melancholic, choleric, and phlegmatic types still usefully describes common behavior types, even if the underlying biological mechanisms that Hippocrates proposed are all wrong.

We gain a certain psychological comfort by sorting people according to types. Once we have labeled a person as a particular type, we can make general predictions about how that person will act in different situations. The ultimate aim of any personality theory is to predict an individual's future behaviors and to explain their past behaviors. Once a type is established, then everybody who fits into a particular category is said to have pretty much the same personality and behavioral predispositions. Many psychological theories have attempted to sort people into convenient types, which include authoritarian, submissive, introverted, extroverted, stable, neurotic, self-actualized, moral, Machiavellian, masculine, feminine, among many more.

Psychologists recognize that although applying personality-type labels is appealing, when you actually try to classify people based on personality types, you immediately run into difficulties. The main problem is that most people are complex and simply don't fit into a single type. In many fictional stories and dramas, the author first leads us to believe that an individual is one personality type and then lets another emerge. Shakespeare made a career out of the presentation of such characters. Were Shylock in *The Merchant of Venice,* Brutus in *Julius Caesar,* and Lady Macbeth truly evil?

To use a more recent but still classical example, consider the fictional character of Vito Corleone, the Mafia don, played by Marlon Brando in the films *The Godfather* and *The Godfather Part II.* What personality type is this character? With Hippocrate's system, we run into immediate difficulties. In some scenes he is clearly depicted as aggressive, willing to have people hurt or killed on his behalf, touchy, sensitive, and easily insulted, quite likely a choleric type. Yet in other scenes he is quiet, a listener rather than a talker, extremely controlled and restrained when others are excited, which might define him as phlegmatic. In still other scenes he is extremely caring toward his children, grandchildren, and wife as in the opening wedding scenes where he demonstrates a confident, optimistic, and cheerful demeanor, all of which suggest a sanguine personality type. While such complexities and complications make good drama, they also frustrate scientists, whose task it is to predict and analyze behaviors. Because of problems similar to that of determining the personality type of the fictional godfather, many modern psychologists have decided that trying to pigeonhole individuals into discrete groupings or types simply doesn't work.

Personality Traits

Personality theory is not the first case where psychologists have had to abandon simply categorizing people into fixed types. They also ran into problems when trying to categorize kinds of intelligence. Initially, there were three intelligence-types: average, genius, and retarded. It became obvious that this classification system was oversimplified and that many individuals show brilliance in certain areas but can appear witless in others. A classic example is Albert Einstein, whose name is virtually synonymous with genius. Einstein was not only a brilliant physicist, but he also had a high degree of verbal intelligence, as his many philosophical writings show. He was also musically talented and played the cello. Yet his intellectual downfall was simple arithmetic. His addition and subtraction skills were so bad that his personal checkbook was always completely out of agreement with the bank records. He called upon his first wife to check the arithmetic in his equations so frequently that some people wondered whether she was really the one responsible for many of his mathematical insights.

We have all heard similar stories about people with unique combinations of diverse skills and deficits, such as the creative and virtually unbeatable chess master who couldn't pass high school geometry. I know an extremely brilliant research chemist who can't follow a simple recipe to bake a cake and a renowned clinical psychologist who can't figure out the first steps for housebreaking his dog. The tendency to act intelligently and to act stupidly can be found in the same person. Whether we are geniuses or dolts seems to depend upon the situation, the problems we have to solve, and the specific abilities required.

Modern intelligence tests measure a broad range of specific abilities and give separate scores for different mental capacities. Typically, these can include measures of the person's vocabulary, mathematical ability, reading ability, creative thinking and problem solving, memory, general knowledge, and many others. People can have a mixture of scores, going from very high on some skills to rather low on others. With a set of such scores, it is no longer possible to classify a person as simply being "smart" or "dumb."

Many psychologists who study personality adopted an approach similar to that used by scientists who study intelligence. They have begun to measure a set of specific aspects of personality. We can't call these aspects abilities, since they are behavioral tendencies (rather than skills), such as moodiness, optimism, aggressiveness, ambition, gregariousness, and so forth. Instead, these behavioral tendencies are called *personality traits.* The real problem in such research is to determine which specific personality traits are important and how many of them we have to consider if we would like to make meaningful predictions about the behaviors of an individual.

At various times some psychological theorists have argued that we must include sixty-four or more personality traits if we want to predict behaviors accurately, while others have argued that we can get enough predictability by just choosing one dominant trait. Over the past quarter century, the research has suggested that we can pretty much describe human personality using just five traits. For this reason, psychologists have named these *The Big Five.* Let's look at these.

- *Extroversion* is a trait that looks at how active, sociable, dominant, or fun-loving a person is. Someone who is low in extroversion is likely to be retiring, passive, not very dominant, and may tend to avoid social encounters.
- *Neuroticism* refers to emotional stability or emotional reactivity. People high in this trait are apt to be moody, anxious, and insecure. They may worry constantly. They typically show frequent changes in their moods. People low on this trait are calm, have few major mood swings, and tend to be more satisfied with life and their own performance.
- *Agreeableness* describes whether the individual is warm and pleasant, as opposed to cold and distant. Individuals who are high in this dimension also tend to be cooperative, courteous, and trusting.
- *Conscientiousness* looks at whether the individual is careful or careless. People high in this trait tend to be dependable, punctual, well-organized, and reliable. Individuals low on this trait tend to be less organized, sloppy in their habits, less productive, often late, and unreliable.
- *Openness* has been the most difficult of the traits for psychologists to agree upon. At one level, it is associated with intelligence, creativity, and imagination. People high in this trait also tend to be more daring, with broader interests. Individuals with low scores for this trait do not appear to be as bright or cultured; they tend to be conforming and to shy away from new experiences.

Do Dogs Have Personalities?

Some psychologists believe that dogs can be described using the same personality characteristics as people. Psychologist Sam Gosling, from the University of Texas at Austin, and his colleagues started out by assuming that animals didn't have personalities. He was interested in determining how human personalities evolved and thought that looking at instances where personality couldn't be presumed to exist, namely nonhuman animals, would be a good starting point. However, he also believed in the theory of evolution, and this ultimately challenged his position.

Like Gosling, my own belief in evolution has often caused me to challenge the notion that humans are unique in terms of important behavioral characteristics and abilities, such as language, consciousness, emotions, and even mathematical ability. Any characteristic that improves the likelihood that individuals and species will survive tends to be passed on genetically, and through gradual evolutionary pressure, such behavioral abilities tend to become more complex and precise.

For any social species, whether humans or dogs, individuals are concerned with two tasks. The first is dominance or control, and the second is sociability or getting along with others. Dominance involves trying to promote yourself in the group, so that you end up with more resources that will keep you alive and ensure that you end up with many offspring. However, it is also important that you get along with other members of your species. Getting along is important because humans must cooperate in many activities to ensure survival. The same is also true of dogs (or at least their wild cousins) who must cooperate in the dangerous task of hunting large animals and the vital tasks of caring for their young. It is also important to avoid conflicts that might injure individuals and reduce the whole group's ability to survive.

Stable personality characteristics and predictable behaviors are actually advantageous for a social group. Thus, if an individual is clever, dominant, and bold, it makes sense to allow that individual to take a leadership position in the social group and not to challenge his every decision. If the individual is meek and fearful, leaving him back at the den to mind the pups puts him in a more helpful position than on the hunt; as a caretaker, his fearfulness is apt to lead him to hide and conceal the young if danger approaches, and this is a behavior that might actually increase survival of the pack. Knowing that a pack member is moody, touchy, and aggressive makes others avoid him, which will reduce physical conflict and potential injuries that could damage the hunting efficiency of the group as whole. Whether you are a person or a dog, it appears that having a personality that can be understood by other members of your social group helps in the efficient functioning of your community.

Any successful evolutionary adaptation or change that has occurred in a species has usually been preceded by some simpler versions in other species. Take, for example, the thumb, a marvelous mechanical device that has given humans the ability to create our technological world. It is an opposable digit, which means that we can touch it to the tips of any of our other fingers and that provides it the ability to manipulate small objects deftly. This skill is vital if we want to create and use tools. This special digit first appeared as a stubby thing, not really opposable against any fingertips, in an ancestor we share with monkeys. During the gradual evolution by which lower primates became apes, the thumb grew longer until in some species the thumb can oppose one or even two fingers to some degree. Clearly, the thumb of humans has evolved from those simpler forms that predated it.

We see the same kind of gradual change in the evolution of flying ability in birds. This was preceded by less complex versions of this skill. For example, some earlier animals, such as the flying dinosaurs known as pterodactyls, could glide through the air. However, their gliding was not true flight. Essentially, they soared through the air with a limited degree of control, much like a person floating through the air on a parachute. The flight ability of birds is just a more advanced and complex evolution of this simpler ability to soar or glide. What has been added in the modern birds is the ability to take off from any surface and to change altitude at will.

Charles Darwin, and those scientists who have come to accept the theory of evolution, believe in the notion that most important and useful abilities show some form of continuous change through history, as species evolve and adapt to their environments. If you choose to ignore the clear evidence of gradual change, then you must accept what biologists call a "hopeful monster" theory. This monster is a miraculous accident in which a freak mutation, just by chance, happens to produce a radically better-equipped animal. This is really the evolutionary theorist's version of "divine intervention."

This kind of explanation, based upon sudden changes that produce unique abilities and behavioral characteristics, makes me, and many other scientists, uncomfortable. Evolution is much like a big highway that all species travel down. Changes in direction are gradual, since too sharp a turn will cause the quickly traveling vehicles (or evolving species) to fly off the road and into extinction. At the biological level, this idea of a highway shows up in the form of continuous slow change, with a lot of similarity between various animal species, especially at the genetic level.

It might be surprising, or even disturbing for some people to learn that there are some recent findings of modern biochemistry that suggest that humans are not as genetically unique as we might have thought. It is well known, for instance, that human and chimpanzee DNA is at least 98 percent identical. This degree of similarity is so close that some scientists have proposed that it might be possible to perform crossbreeding and to make a hybrid species, although, of course, one would hope that moral and ethical considerations would forbid such a genetic experiment. Even an animal that looks very different from humans, such as our pet dog, is molecularly and genetically quite similar to us. We are both mammals and the DNA codes of dogs and humans have in excess of 90 percent agreement with each other.

Since we are so close genetically to nonhuman animals in all other aspects, it seems unlikely that evolution would make a sudden quantitative and qualitative jump in the development of personality. Darwin noticed that emotional responses, which are part of the package of behaviors that personality predicts, are similar in animals and humans. It seems likely that if evolution was on a highway headed toward the appearance of human-level personality characteristics, these should first make their appearance in the behavior patterns of simpler social animals—such as dogs. The logical expectation is that the personality of animals such as dogs will not be as complex as the personality of people, perhaps with fewer identifiable personality traits, or a lower level of consistency, but the logical conclusion is that there will be an identifiable personality in dogs.

Of course, humans clearly differ from animals in many ways. For example, humans are the only animals that put on bathing suits before going into the water, pierce body parts and insert ornaments into the holes, dye their hair, tattoo their bodies, paint their faces, use money, and cook food. While such differences may appear important, they actually have little bearing on our success or survival as a species. How such behaviors came about is unknown, but because these behaviors have such a minimal influence on survival, we would call them cultural rather than evolutionary. Personality certainly does have survival value. For domestic dogs, whose environment is now the world of men, personality is the factor that most determines their survival.

Today, as would have happened during the process of domestication, a vicious, uncooperative, unsociable animal would not survive in the world of men. It would be killed for the safety of our children and the community, and would not be protected, fed, or encouraged to breed. In Belyaev's domestication of foxes, his sole selection rule for breeding was a personality characteristic that included sociability or friendliness. That sounds much like selection for the Big Five personality trait of agreeableness. If we are selecting animals based upon a personality trait similar to that found in people, then, at the very least, it would appear that dogs have at least one personality trait in common with humans. Sam Gosling and his research associates would contend that there is a lot more similarity than that.

Dogs and the Big Four

In a landmark study published in the *Journal of Personality and Social Psychology*, Gosling and his collaborators started out with the idea that the human five-factor theory of personality could be applied to dogs as well as humans. He used the same testing and analysis procedures that psychologists employ to measure human personality and modified them so that they could be applied to dogs. The results of his work on dogs suggest that their personality structure, while somewhat similar to ours, may be simpler because it contains only four traits, rather than the five needed for people. The missing fifth trait in dogs is conscientiousness. Although we might see some aspects of this personality trait in dogs, such as the ability to focus on a task, the vast array of behaviors that define this trait—such as organization, orderliness, a sense of time and sequence, a sense of ethics, purpose, and punctuality—don't seem to apply to canine behaviors. As Gosling explained in an interview, "In our previous research we found there's very little evidence for a separate conscientiousness dimension in any other species than humans and chimpanzees." Judging by the absence of neatness and discipline in the spontaneous behaviors of most dogs that I have lived with, I agree with him and believe many other dog owners will too.

The initial problem in Gosling's research was to decide what the dog equivalents to the remaining four dimensions (extroversion, neuroticism, agreeableness, and openness) were. This involved specifying the kinds of behaviors that are characteristic of each personality trait that can be objectively observed in dogs. In the canine version of personality, extroversion becomes energy level, emotional stability becomes anxiety level, openness becomes intelligence, and agreeableness becomes affection and sociability.

Gosling's research involved seventy-eight dogs and their owners, plus seventy-eight other people who knew the dogs, and a set of strangers who did not know the dogs but would observe them. In the first stage of the research, owners and the people who knew the dog described his or her personality. In human psychology, researchers sometimes use descriptions provided by friends, relatives, or a variety of "significant others" who know a person, in order to gain some insight into an individual's personality. Judgments of the dogs' personality were collected with a standard questionnaire that researchers often use when testing humans. Obviously, some items had to be adapted slightly so that they applied to dogs. Thus, an item like "Is original, comes up with new ideas" would be changed to "Is original, comes up with new ways of doing things." Only one item in the entire inventory designed for measuring human personality could not be reasonably translated to a canine form, namely "Has few artistic interests," so it was omitted from the test.

The dogs' owners and other people who knew the dog judged how well each item in the questionnaire described the dog by using a five-point scale ranging from strongly agreeing to strongly disagreeing that the dog had each particular characteristic. A typical item for extroversion/energy would be "Is full of energy." Agreeableness/affection was measured with items like "Is cooperative." Neuroticism/emotional reactivity was judged with descriptions like "Can be tense," while openness/intelligence was determined using descriptions such as "Is curious about many different things." None of the people reported any difficulty describing the "typical behaviors" of the dogs, which suggests that, in their experience, each dog acted as if it had a personality and predictable behavior patterns that distinguished it from other dogs.

In the last stage of testing, three strangers observed the dogs as they were being tested in a fenced region of a park. The idea was to get the animals to perform a number of different activities to see if the personality judgments made by the people who knew the dogs actually predicted what the dogs would do. This field test began with a short "getting acquainted" walk, with the dog and owner accompanied by the strangers who had watched the animal's behavior. Energy level was assessed when owners encouraged the dogs to be active by running and playing. Obedience was measured by the dog's response to basic commands such as "sit" and "stay." A problem-solving test involved timing how long it took for a dog to find a treat placed under a plastic cup. The anxiety tests monitored the dog's emotional reaction when a stranger took him for a short walk and then again when his owner walked away with a strange dog. Finally, in the affection test, owners encouraged their dogs to be snuggly and show affectionate behaviors.

Gosling confirmed that dogs had reliable, predictable personalities for the four personality traits he measured. Furthermore, there was a lot of agreement between the way that the owners described their dogs' behaviors and the descriptions from both sets of observers, those familiar with the dog and those who met the dog for the first time when the test began. This suggests that global assessments of a dog's personality work pretty much the same way they do with people: dogs seem to project their personality through their behaviors in a manner that others can read and respond to.

Remember that the evolutionary value of having an interpretable personality is to improve the functioning of the social group by allowing others in the group to predict that individual's responses to particular situations. If they can anticipate the behavior of another individual, then they can plan their own responses and act in a manner that will avoid difficulties and conflicts. As Gosling expected, the results of the personality tests predicted much the same pattern of behaviors that the dog actually showed during the field testing.

Dogs have fewer personality traits than people do, which is consistent with the idea that their mental structure is simpler. It is also possible to ask if the personality traits of dogs are less well defined than those applied to people. If this is the case, then dogs would not be as consistent in their behaviors and our judgments of a dog's personality would not be as reliable a predictor of behavior as our judgments of a human's personality. When asked, "How strong was the agreement between the personality based on familiarity with the dog and the personality based on actual tested behaviors?" Gosling answered, "About as strong as you would expect from the ratings of humans, and that's the essential point—all of the tests for personality judgments of dogs met the criteria that we expect for personality judgments of humans. So, if you want, you could say, 'Well, dogs don't have personality.' But if you do say that, you're also forced to say that humans don't have personality either."

Gosling also showed that, even though breed is a reasonable predictor of temperament (an issue that becomes more important later on in this book), personality varies widely within a breed, thus, not all German shepherds are clever; not all rottweilers are protective; nor are all golden retrievers affectionate.

Do people use the same words to describe the personalities of dogs and people? Sure they do—as when they describe a dog as aggressive, clever, friendly, or shy. However, I recently asked a woman to describe the personality of her golden retriever, and she responded with "He's just one big hairy smile." In all my years as a psychologist, I've never heard a human being described in that particular way!

CHAPTER FIVE

The Personalities of Dog Breeds

It had been a cool damp day in Canada's southern Alberta. The retriever trials had run late, but most of the dogs, ducks, and equipment were now safely stowed away to wait for the next day when competition would continue. A collection of six or seven old and new friends had gathered in the warmth of a motel room to chat. A few dogs lay nearby, and an open bottle of bourbon and another of Canadian whiskey sat on the table.

It was a typical gathering of sporting dog owners, and the conversation, as usual, consisted of reminiscences, training advice, and many humorous stories or quips. As the alcohol warmed and relaxed us, we became a bit more philosophical. We had just been discussing the issue of whether dogs went to heaven, and the conversation had moved from serious to hilarious as we speculated on what modifications and facilities would be needed to accommodate our pets in the hereafter.

It was then that Ralph, a longtime field dog competitor, leaned over and put his grizzled hand on the head of his nine-year-old golden retriever, Morgan.

"Well, when Morgan, here, gets to heaven, the first thing he'll do is to ask to see Saint Peter. Next he'll ask to meet Jesus, and then he'll ask to meet God and the Holy Mother—you see, for golden retrievers meeting people *is* heaven!"

The members of the group laughed or nodded and the conversation began to drift to another topic. The humor in Ralph's suggestion lies in the fact that he was talking about a golden retriever, and his comments fit with the conclusions that most people have about the personality of golden retrievers. His comments would have made no sense if he were talking about a Rottweiler, Doberman pinscher, or Akita, and certainly not a wolf.

The difference in sociability between dogs and wolves confirms that genetics help shape canine personality, however, the best examples of how genetics contributes to your dog's personality and behavior are the many differences among dog breeds. By definition, each breed is the result of selective breeding that deliberately manipulated the genetic makeup of the representatives of that breed. The studbooks and pedigrees of dogs of each breed trace the genetic heritage of each individual dog. Breeders will proudly tell you how accurately their line of dogs captures the physical and temperament characteristics that define the "standard" or ideal dog of their breed.

Every dog breed is different, in not only size, color, and shape, but also in their instinctive, inherited behavioral tendencies, which includes a personality profile that is specific to each breed. Most people recognize certain aspects of these breed differences, and these even form the basis of a never-ending piece of humor that starts with the question "How many dogs does it take to change a lightbulb?" The answer depends on the breed.

> ***Border collie:*** Just one, but why not let me change the light fixture so that it will accept fluorescent bulbs, which are much more efficient?
> ***Bulldog:*** Don't bother; I'll just lie here in the dark.
> ***German shepherd:*** Just one, but it will have to wait until I've rescued those people trapped in the dark and led them to safety, and then checked the house to make sure that no one has entered under cover of darkness to take advantage of the situation.
> ***Shetland sheepdog:*** I will as soon as I arrange all of the new bulbs in a tight and orderly little circle.
> ***Golden retriever:*** The sun is shining, we've got no work to

do today, I've got this neat red ball here, and you're inside worrying about some silly lightbulb?

Rottweiler: Go ahead, and see if you're tough enough to make me!

Corgi: First, I'll bark until the old bulb leaves of its own accord and then I'll nip at the new one until it goes into the socket . . .

Labrador retriever: I can do it! Please . . . Please . . . I can. You know I can. Please . . .

Greyhound: It isn't moving, so who cares?

Pointer: I see it. There it is. Look, it's right there . . .

Jack Russell terrier: Me! I can reach it! All I have to do is to keep bouncing off the furniture and walls.

Poodle: I'll just whisper some sweet nothings into the border collie's ear and he'll do it. By the time he's finished putting in the new light fixture, my nails will be dry.

Obviously, the list can go on to include many other breeds. The humor in this obviously comes from the fact that most people have strong beliefs about the nature of each breed's personality and behavioral tendencies.

A New Rating of Breed Differences in Personality

Several scientific studies have used the opinions of people who are experts on dogs and dog behavior to try to determine the traits that make up the personalities of different dog breeds. Global assessments that simply ask, "Is a dog of that breed friendly?" don't work as well as do measures of specific observable behaviors. Thus, a researcher might start by isolating commonly observed behaviors that might reflect personality differences in dogs, much as Gosling did in his field tests. The next step is to survey professionals who work with dogs and have presumably had opportunities to observe these behaviors in various breeds.

One well-respected study of this sort was conducted in 1985 by

Benjamin and Lynette Hart, both of whom are now with the Center for Animal Behavior at the University of California at Davis. They surveyed forty-eight dog obedience judges and forty-eight veterinarians who rated the fifty-six most popular dog breeds in that year (according to American Kennel Club registrations). Each expert was asked to rate the dog breeds on thirteen different personality characteristics that the Harts thought were most useful, including: excitability, general activity, snapping at children, excessive barking, playfulness, trainability, watchdog barking, aggression to dogs, dominance over its owner, territorial defense, demand for attention, destructiveness, and ease of housebreaking. An example was given to make clear what each characteristic represented. For instance, data on excitability was asked for in the following way: "A dog may normally be quite calm but can become very excitable when set off by such things as a ringing doorbell or an owner's movement toward a door. This characteristic may be very annoying to some people. Rank these breeds from least to most excitable." There was generally a good degree of reliability and agreement on rating the various breeds on most of the behavioral tendencies, although the agreement on a few characteristics, such as ease of housebreaking, destructiveness, and demand for attention was low.

The first problem with this method of looking at the personality of dogs is that the characteristic behaviors that you choose might not be those that are most important or have the most influence on people's interactions and satisfaction with their dogs. For this book I decided to revisit the question of the personality of particular dog breeds and started, much the way the Harts did, by making a list of what I thought were the most important characteristics. I began with the dog's activity level, intelligence, friendliness, and so forth. It quickly became clear to me that my list might be very idiosyncratic, reflecting the fact that I live in a city and keep my dogs mostly as companions. The job that my dogs have, if they can be said to have one, is to compete in obedience competitions and to serve as demonstration dogs in my classes to show students what dogs can actually learn. Yet if I wanted to assemble a good general picture of the important characteristics of dogs, I needed some other viewpoints, so I contacted a number of dog experts. These included two veterinarians, two dog obedience instructors, two dog obedience judges, two conformation, or show dog, judges, a field dog trial judge, a tracking trial judge, and a dog behavior analyst. All of these people knew a lot about dogs, so I asked them to make a list of all of the personality, behavior, and temperament characteristics that they thought might be important in determining how well a dog could get along in our human world.

Next I took the twelve lists (eleven from my group of experts plus my own), and sorted through them. These lists included just about every imaginable dimension of dog behavior. To my experts, activity level was important, but they also distinguished between indoor and outdoor activity, general restlessness, and even the speed at which the dog grabs food from his owner's hand. In the eyes of these experts, sociability and dominance are not unitary dimensions but each had to be subdivided into how friendly or dominant a dog was in the presence of different individuals. For example, there were separate listings for the dog's sociability (and dominance) around adults, with children, within the family, with strange adults and unfamiliar children. The experts even included sociability around familiar dogs, unfamiliar dogs, and other animals.

Ultimately, I was able to reduce the number of behaviors to what appeared to be the most critical twenty-two dimensions of dog behavior. To be on the final list a characteristic had to be mentioned by more than half of my experts, and it had to be a dimension that other dog experts could reasonably evaluate for any dog. These characteristics included dominance, territoriality, guard dog ability, watchdog ability, friendliness to strangers, congeniality toward family and people who are well known, amicability around children, indoor activity level, outdoor activity level, suitability for city living, exercise needs, learning ability, problem-solving ability, suitability for obedience training, willingness to work for people, how vigorous, determined, and persistent the dog is, gentleness in human interactions, the predictability or constancy of the dog's day-to-day behaviors, and how emotionally stable or constant the dog is.

I next contacted one hundred fifty different dog specialists, including veterinarians, dog trainers of several sorts (basic obedience, competition training, agility and other dog sport trainers, protection dog trainers, et cetera), dog judges of all kinds (obedience, conformation, field trials, search and rescue, tracking, terrier trials, et cetera), several writers of dog books, some canine psychologists and behavior analysts. I gave each of them a list of dog breeds and asked them to rate each of the breeds that they felt they knew well on each of the twenty-two dimensions. Since I knew that this would involve a major commitment of time, I gave each expert only forty breeds. Each expert had a different list of forty breeds, so that all of the major breeds would eventually be covered. In addition, the experts could add any breeds that they felt that they knew particularly well and wanted to rate, or drop any breeds that they didn't know well enough to rate.

I became a bit concerned after sending out these survey forms to the other dog experts, because when I sat down and did this task myself, it took me a couple hours to rate my own forty breeds. I became concerned that not many experts would actually take the time to complete this chore, but I was pleasantly surprised to find that ninety-seven busy professionals took the time to return a set of ratings to me. My minimum requirement for a breed to be included in the study was that I had full ratings from at least twenty-five of the experts on the twenty-two behavior dimensions. When I looked at the data, I found that 133 breeds of dogs reached that criterion for inclusion. Actually, for most of the breeds I had close to fifty ratings, since virtually all of the experts included additional breeds that they knew well and wanted to rate. One dog obedience judge went so far as to send me a complete set of ratings on 108 breeds of dogs, which must have taken him a full day to complete.

Finding Personality Traits That Matter

For the purposes of discussing the personality of dog breeds, my first task was to reduce all of the listed dimensions to a manageable number since it was likely that some of these behavioral dimensions were not truly distinct but were measuring different aspects of the same basic psychological characteristics. To do this, I used a sophisticated statistical procedure called *factor analysis* which looks at how data cluster together and can give us a picture of underlying psychological processes. I was lucky and found that all the data could be reduced to five understandable personality characteristics. Four of these roughly agreed with the personality traits discussed in the previous chapter, which Sam Gosling found were relatively the same in humans and dogs. The first was a measure of energy level, which is most related to activity levels indoors and out, and is somewhat similar to an extroversion/energy dimension in people. The next reflected learning, obedience-training ability, and problem solving and is similar to the openness/intelligence measure for humans. Another was sociability, which measures how well the dog gets along with people and other animals and is approximately the equivalent of agreeableness/affection in people. The fourth was stability versus excitability, which measures, among other things, how quickly moods or emotional states change and how large the emotional swings are, which is equivalent to the human dimension that we call neuroticism/emotional reactivity.

The fifth dimension, which did not appear in Gosling's work but was quite clear in the ratings that my group of experts gave to me, looked like a measure of dominance and territoriality. Many characteristics that involve aggressive behaviors are related to this dimension. This added dimension was interesting but not completely unexpected. In 2002 two ethologists, Kenth Svartberg, at Stockholm University, and Bjorn Forkman, at the Royal Veterinary and Agricultural University in Frederiksberg, Denmark, analyzed objective behavior test results from more than fifteen thousand dogs. Their data came from the Swedish Working Dogs Association (SWDA), which was involved in the selection of service dogs—specifically police dogs, explosive and drug detection dogs, search and rescue dogs, guide dogs for the blind, hearing assistance dogs, and so forth. When they mathematically analyzed their data, they also found that dogs' personality structure could be described by five basic personality traits that were generally similar to my findings and Gosling's. Theirs were sociability (agreeableness), curiosity versus fearfulness (which seems to combine both the intelligence and emotional reactivity measures), plus two other dimensions that they called playfulness and instinct to chase (which together would probably be related to extroversion and energy level in my measures). In addition, however, they found a separate dimension that they called "aggressiveness" which seems very close to what I call dominance/territoriality. This is reassuring, since it shows that different ways of testing canine personality characteristics produce generally similar results. Such consistency is always heartening to a scientist.

Reclassifying Dog Breeds

Now that we have the five basic dimensions of dog personality, we can start to look at the personalities of the individual breeds. To discuss each breed separately and in detail would take an entire book on its own, but I have provided a table in the appendix that contains the rankings of each of the 133 dog breeds that the experts rated on the five basic personality dimensions. But before you rush to the end of the book to see how your favorites fared, please consider the following overview which shows some interesting results about the different personality profiles of breeds.

In order to get a good picture of the personality of purebred dogs, we first have to be able to group the dogs in some way. From the time of the Renaissance, people interested in dogs have found it convenient to classify breeds into broad groupings according to their functions. Early authorities used fairly large categories, such as "beast dog" (for hunting dogs), "coach dog" (which included guard dogs), and "vermin dog" (containing mostly the terriers). In England, one of the earliest classification schemes reflected the popular prejudice against immigrants at that time by including a category called "foreign dog" as a separate class. With the founding of the Kennel Club of England in 1873, a more formal breed classification was introduced and, with minor modifications, the American and Canadian kennel clubs adopted systems similar to the one established by the British club.

We could use the American Kennel Club (AKC) groupings as a starting point, but you will see that we have to expand them if we want to create meaningful groupings of personality and behavior. Why should we use the AKC groupings at all if we have to modify them? Mainly because they are so well known. If the AKC were a software development firm, it would be the equivalent of Microsoft—omnipresent and dominant. Even if you don't use their systems, you certainly would know about them.

It is important to remember that the kennel clubs group dog

breeds based on their functions. Not surprisingly, the first kennel club grouping is *sporting dogs* in the United States and is called *gun dogs*, by the British. These dogs are supposed to assist a hunter. In this group are breeds that find game by scent, including some that indicate the presence of their quarry by pointing to it and some that flush or spring the game so that it can be shot by a person with a rifle or shotgun. This group also includes dogs specifically designed to reduce the labor and discomfort associated with hunting by retrieving game both on land and from the water. The list of sporting dogs includes some of the most recognizable breeds, such as the spaniels and retrievers. Also, in this group are two breeds, the Labrador and golden retrievers, who are among the most commonly used service dogs, assisting blind or handicapped people, and are also used to detect drugs and explosives and assist in search and rescue tasks.

The sporting group actually contains five different types of dog, each of which has somewhat different purposes and thus might be expected to have different personalities. These classes of dogs are the *retrievers*, who must keep in continual contact with the hunter and take direction from them; *pointers*, who work steadily and quietly, pointing to the game and waiting for a signal to be released; *spaniels*, who quarter the field and flush game at will; and *setters*, a faster working dog that, like the pointer, is designed to locate game; plus some *multipurpose* dogs (vizsla and Weimaraner) that are designed to do a combination of tasks like pointing, flushing, and retrieving game.

The *hounds*, which make up the second AKC grouping, are also hunting dogs, but of a much more independent variety. This is a very diverse group, and includes the fastest runners in dogdom, such as the greyhound, Afghan hound, and the Saluki. It also includes dogs that poke along even when going at full speed, such as the basset hound and dachshund. In addition, this group contains dogs that were specially bred to hunt particular kinds of game, including the Irish wolfhound, Scottish deerhound, otterhound, Norwegian elkhound, foxhound, and coonhound. As you can see, all of these contain the name of their quarry in their own breed label. Included in this group is the dachshund, since *dachs* is the German word for badger, which was the game that these little dogs, with their powerful jaws, were supposed to hunt. One characteristic of all hounds is that they are easily distracted by sights and scents and are apt simply to wander off when they encounter something interesting.

Hounds naturally fall into two clear groupings, plus three breeds that, based on behavior and other characteristics, belong somewhere else. The first group is the *scent hounds*—dogs that are supposed to track their quarry by the faint odor they leave as they move over the landscape. The second group, the *sight hounds,* has keen eyesight and tremendous speed. Their task is to visually locate their quarry in the distance and run it down. The hunter is only needed if the game finds refuge in a den, burrow, or tree. In all other cases, these hounds are supposed to be able to dispatch their target when they catch it, without any human intervention.

Some hound breeds are better sorted into other groups. The dachshund, for instance, was designed to dig into burrows after its quarry, so it arguably belongs with the vermin-seeking terriers. The Rhodesian ridgeback was originally designed to hunt lions and hold them at bay, but actually was more frequently used as a personal protection guard dog and thus we will put it into that group. Finally, the Norwegian elkhound is a spitz-type dog, a group that includes all of the northern-type dogs that people often call huskies. These dogs have pricked ears, sharp muzzles, a flowing tail that curls high over their backs, and often dense insulating coats.

The *working dog* group has various practical functions, such as guarding or protecting property or people, pulling or hauling loads of materials, or doing search and rescue work. All of these tasks require strong, muscular dogs that range in size from medium to large. Familiar examples are popular guard dogs, such as the Doberman pinscher and Rottweiler. Dogs designed for pulling sleds or carts, or for general rescue work include the Alaskan malamute, Siberian husky, Bernese mountain dog, Great Pyrenees, Saint Bernard, and Newfoundland.

The working dogs naturally divide into four groups. The first is the *guard dogs,* designed to protect livestock or territories without man's direction, such as the Great Pyrenees and the kuvasz. The second group includes *personal protection dogs* that were bred to work under direct human control, such as the Doberman pinscher and boxer. The third group includes the *draft dogs,* bred to pull carts and carry packs, such as the Bernese mountain dog and the Newfoundland. We separate these out from the dogs that pull sleds, since those are Nordic breeds or *spitz-type dogs,* such as the Siberian husky or malamute, which make up their own group based on their behavior and temperament. One breed that does not belong in the working dog group is the Portuguese water dog, which is really a multipurpose sporting dog that has been successful as a water retriever and is as good a rabbit hunting dog as many field dogs.

The AKC *terrier* group includes a broad range of big and small dogs designed to keep farms and farm buildings free of vermin, such as rats or foxes. The word *terrier* comes from the Latin root *terra,* meaning earth or ground, which provides a hint about how these dogs frequently hunt their prey. Terriers may be required to dig underground or wriggle their way into burrows or dens to find their targets. Because of this, many were kept deliberately small. Some were also bred to have rough, hard, or wiry coats that protect them against abrasion from rocks and rough ground, and provide a kind of armor against the teeth of their quarry. Some of these small, short-legged terriers were carried in a basket on horseback during hunts. They were only put to work when the game animal had been driven into its den. Other, larger terriers with long straight legs were designed to run with horses and hounds on the hunt. They could also pursue fast-moving vermin over open ground. The names of terriers often reflect the geographical location where they were first bred, such as the Australian terrier, Airedale terrier, Scottish terrier, Irish terrier, Manchester terrier, Welsh terrier, Kerry blue terrier, and West Highland white terrier. Some others carry names of individuals who were supposedly important in their development, such as the Parson Jack Russell terrier, or the Dandie Dinmont terrier. Together this makes one large group of *vermin-hunting dogs,* but the terriers also include the smaller grouping of *fighting dog breeds,* such as the Staffordshire and American Staffordshire bull terriers.

The AKC *toy* dog group contains dogs bred specifically to provide companionship. These were supposed to be small, lightweight, portable, and agreeable—the sort of pet that aristocratic ladies could carry with them. Some of these dogs were originally owned by royalty, such as the Pekingese. No one outside of the Chinese royal families was allowed to own these, and any attempts to take one of these dogs out of their country of origin was punishable by death. Such dogs were often prized more for their look than for any other characteristic. Thus, the Pekingese was supposed to look like the Chinese celestial lion, and each year a contest was held to determine the dog that best approximated this ideal. Other toy dogs were designed to be smaller versions of familiar large dogs. The Cavalier King Charles spaniel and the papillon are miniature spaniels, the Italian greyhound is a miniature version of the larger greyhound, and the toy poodle is a scaled-down version of the standard poodle. Other dogs found in this group include tiny terriers, like the Yorkshire terrier and the toy Manchester terrier, plus some dogs with fancy coats, like the shih tzu or Pomeranian, or no coat at all, like the Chinese crested. In North America size seems to be the defining characteristic for this group, though in Europe these dogs are defined by their function and called *companion dogs,* which allows the inclusion of some larger breeds with similar temperaments such as the English bulldog.

The AKC has separated the *herding dogs* from the other working dogs. This group contains dogs that were used, at least originally, to herd, drive, and protect livestock. The herding group includes varieties of collies, corgis and Belgian shepherds, plus the familiar Old English sheepdog and the German shepherd, and slightly more exotic dogs such as the puli and the briard, among others. In Britain, these dogs are still considered part of the working dog group.

There are two types of herding dogs. The first are predominantly flock *herding dogs,* indispensable in their ability to keep herds of animals together and to move them short distances under the direction of a human shepherd. Examples of this group would be the border collie, German shepherd, and the Shetland sheepdog. The second type of herding dog is the *drover.* These hardy dogs were bred to drive sheep and cattle over long distances, often with minimal or no direction from humans for long periods. The Australian cattle dog, Bouvier des Flandres, and corgis are in this group.

I saved the AKC *non-sporting* group for last since it is a potpourri of dogs that appear to have been included in this group simply because the AKC felt they didn't fit into any of the other classes. In

Britain, there is no comparable catchall classification. Most of the non-sporting breeds are recognized there, but they are distributed throughout the other groups. An argument could be made that some of these breeds belong in some of the other groups. For example, a case could be made that the Dalmatian, Chinese shar-pei, and schipperke should be in the working group since all have served as watchdogs—the Dalmatian for carriages, the shar-pei for temples, and schipperke for barges. The poodle could be put in the sporting group, since it started out as a retriever, and even today, does well in field trials for retrievers. Similarly, the bichon frise, Lhasa apso, shih tzu, Tibetan spaniel (which is not a true spaniel), and the Tibetan terrier (which is not true terrier), could probably be placed in the toy group as all were designed as small companion dogs, while bulldogs are simply heavier and larger companion dogs. The chow chow belongs in the spitz dog group.

When we look at this revised classification scheme, there are sixteen groups of dogs, namely:

Retrievers: Chesapeake Bay retriever, curly-coated retriever, flat-coated retriever, Golden retriever, Labrador retriever, miniature poodle, Nova Scotia duck tolling retriever, standard poodle.

Pointers: Brittany spaniel, German short-haired pointer, German wire-haired pointer, pointer, wire-haired pointing griffon.

Spaniels: American water spaniel, Clumber spaniel, cocker spaniel, English cocker spaniel, English springer spaniel, field spaniel, Irish water spaniel, Sussex spaniel, Welsh springer spaniel.

Setters: English setter, Gordon setter, Irish setter.

Multipurpose Sporting Dogs: Portuguese water dog, vizsla, Weimaraner.

Sight Hounds: Afghan hound, basenji, borzoi, greyhound, Irish wolfhound, saluki, Scottish deerhound, whippet.

Scent Hounds: American and English foxhounds, basset hound, beagle, black and tan coonhound, bloodhound, harrier, otterhound.

Guard Dogs: bullmastiff, Dalmatian, Great Dane, Great Pyrenees, komondor, kuvasz, mastiff, Rottweiler.
Personal Protection Dogs: boxer, Doberman pinscher, giant schnauzer, Rhodesian ridgeback, standard schnauzer.
Draft Dogs: Bernese mountain dog, Newfoundland, Saint Bernard.
Vermin Hunters: Airedale terrier, Australian terrier, Bedlington terrier, border terrier, Boston terrier, cairn terrier, dachshund, Dandie Dinmont terrier, Irish terrier, Kerry blue terrier, Lakeland terrier, Manchester terrier, miniature schnauzer, Norfolk terrier, Norwich terrier, Parson Jack Russell terrier, Scottish terrier, Sealyham terrier, soft-coated wheaten terrier, silky terrier, skye terrier, smooth-coated fox terrier, Welsh terrier, wire-haired fox terrier, West Highland white terrier.
Fighting Dogs: American Staffordshire bull terrier, bull terrier, Staffordshire bull terrier.
Companion Dogs: affenpinscher, bichon frise, Brussels griffon, bulldog, Cavalier King Charles spaniel, Chihuahua, English toy spaniel, French bulldog, Italian greyhound, Japanese chin, Lhasa apso, Maltese, miniature pinscher, papillon, Pekingese, Pomeranian, pug, shih tzu, Tibetan spaniel, Tibetan terrier, toy poodle, Yorkshire terrier
Spitz-Type Dogs: Akita, Alaskan malamute, chow chow, keeshond, Norwegian elkhound, Samoyed, schipperke, Siberian husky.
Herding Dogs: Australian shepherd, bearded collie, Belgian Malinois, Belgian shepherd, Belgian Tervuren, border collie, collie, German shepherd dog, Old English sheepdog, puli, Shetland sheepdog.
Drovers: Australian cattle dog, Bouvier des Flandres, briard, Cardigan Welsh corgi, Pembroke corgi.

What about mixed breed, mongrels, or what are sometimes referred to as "random-bred" dogs? A dog's breed is determined by its genetic makeup. The particular collection of genes that defines a breed allows us to predict its behavior as well as its size, shape, and coat color. When we crossbreed, we lose some of that predictability, since which genes will be passed on by each parent and how they will combine is a matter of chance. Fortunately, there is some data to suggest that we can still make predictions of a mixed breed dog's personality and behavioral predispositions without knowing much about its parentage. John Paul Scott and John L. Fuller carried out a series of selective breeding experiments at the Jackson Laboratories in Bar Harbor, Maine. By a happy chance, their results revealed a simple rule that seems to work. Their general conclusion was that a mixed breed dog is most likely to act like the breed that it most looks like. Thus, if a Labrador retriever–German shepherd cross looks much like a Labrador retriever, it will probably act much like a Labrador retriever. If it looks more like a German shepherd, its behavior will be very German shepherd–like. On the other hand, as we discussed earlier, most mixed-breed dogs have some predispositions and behaviors that are characteristic of both breeds that contributed to it. The more of a blend that the dog's physical appearance seems to be, the more likely that the dog's behavior will be a blend of the two parent breeds. Thus, if you want an estimate of a mixed-breed dog's personality, first decide which pure breed it looks most like and then use that as your prediction. It won't be 100 percent accurate, but it should be close.

Sometimes the best that you can do is to use the breed groupings that we have described. While it might be difficult to say exactly what breeds went into a particular dog, it is easier to classify it as looking like a typical member of a group, such as spaniel or companion dog. With that information and the data that follow, you can make a fairly good estimate of a mixed-breed dog's personality predispositions.

I have often experienced the value of using the breed groupings to predict a dog's behavior such as when my daughter by marriage, Kari, rescued a puppy that she named Bella. All that could be said about Bella's breeding is that she was shaped very much like a sight hound, although smaller than most and with the wrong tail shape and posture. I have no idea which specific breeds (or mixes) mated to produce her (although we have often entertained ourselves by debating that question). However, consistent with her general shape and appearance, she has the temperament, emotional stability, activity level, and training problems that one expects in a sight hound. Although I still have not much of a clue as to her heritage, it is relatively easy to guess how she will behave in most situations by simply treating her as a sight hound. On rare occasions, she will surprise me, such as the day she hurled herself into a stream to retrieve a toy, since most sight hounds do not like water or swimming. As one dog behaviorist told me, "You can always predict that random-bred dogs will occasionally produce random behaviors."

Intelligence/Learning Ability

The experts' ratings of dogs on the various personality traits were converted to quartiles, or 25 percent groupings. This means that *very low* is the lowest 25 percent compared to other dogs ranked for each trait (75 percent of all dogs score higher than this); *moderately low* is higher than the lowest 25 percent of the dogs but lower than 50 percent of the dogs ranked for each trait; *moderately high* is higher than 50 percent of the dogs but lower than the top 25 percent of dogs ranked for each trait; *very high* is the highest 25 percent compared to other dogs ranked for this trait (75 percent of all dogs score lower than this).

The personality data show systematic differences in the personality profiles of the various groups. Let's start with the intelligence dimension (the trait labeled openness in humans). This is a measure of learning and obedience ability but also includes problem-solving ability. Dogs that score highly on this dimension should be relatively easy to train for things like basic obedience commands and service work. Based upon the ratings of my sample of experts we get the following pattern of rankings for intelligence and learning ability. (Note: when breed groups are at the same approximate level, the groups that rank highest are listed first.)

Very High Learning Ability: herding dogs, retrievers, drovers, personal protection dogs.
Moderately High Learning Ability: draft dogs, pointers, spaniels, multipurpose sporting dogs.
Moderately Low Learning Ability: vermin hunters, setters, spitz, companion dogs.

Very Low Learning Ability: fighting dogs, guard dogs, sight hounds, scent hounds.

Thus, herding dogs and retrievers top the list in intelligence, which probably explains why the top-ten lists of dogs in obedience competitions tend to be dominated by border collies, Shetland sheepdogs, and Labrador and golden retrievers. This is also consistent with the fact that German shepherds, Labrador retrievers, and golden retrievers are frequently the preferred breeds for service dogs. Although the breed groups predict intelligence quite well, there is variability within them. For example, the vermin-hunting dogs include border terriers, whose learning ability is ranked in the top 25 percent for all dogs, and the Scottish terrier, ranked with the lowest 25 percent.

Don't immediately rush out to purchase a dog that is in one of the highest ranks for intelligence, nor contemplate giving away your lower-ranked dog. While a dog that is high in learning ability is good to have if you are going to be asking it to perform many duties, there is a downside to having such a smart dog. Such an animal not only learns everything you want to teach him, he can also learn what he can get away with, as well as many things you had no intention to teach him. One woman who owns a Labrador retriever told me that her dog watched her young daughter playing in the kitchen, opening cabinets and drawers. From that simple observation of the child's activities, the dog learned to open cabinets to get at dog food stored there, and to open drawers where various dog treats were stored. These were not skills the owner found desirable.

Sociability

Sociability measures how agreeable dogs are around people. Dogs high on this dimension seek attention, want to be with people, and are friendly, affectionate dogs. Dogs low on this dimension are often described as aloof or standoffish. Highly sociable dogs are seldom shy while dogs low in sociability are often shy and may even avoid human contact, especially with unfamiliar people. In a human personality, this trait is usually labeled *agreeableness.*

Very High Sociability: setters, scent hounds, retrievers, spaniels.
Moderately High Sociability: multipurpose sporting dogs, draft dogs, pointers, spitz.
Moderately Low Sociability: herding dogs, vermin hunters, guard dogs, companion dogs.
Very Low Sociability: drovers, personal protection dogs, fighting dogs, sight hounds.

The experts' opinions seem to confirm the popular stereotype that setters, retrievers, and spaniels are extremely friendly, loving dogs. It may be a bit of a surprise to find that the scent hounds are so friendly as well, although as the owner of a beagle, I can say that his desire for social contact is extreme and, in some instances, such as when I am trying to do some work, can be intrusive. However, such dogs are overall a pleasure to have around. It is not surprising to find that the personal protection dogs and fighting dogs score so low on this dimension. The sight hounds are also low but make up one of the most variable groups in sociability, with the Scottish deerhound scoring in the highest 25 percent of dogs for sociability while the saluki scores in the lowest 25 percent.

Energy Level

Next, let's consider energy level. In a human personality, this is most closely related to *extroversion* which is generally viewed as the opposite of *introversion.* In dogs, this is actually a combination of both indoor and outdoor activity levels, which really can be quite different. For example, the greyhound loves to run at full speed when outdoors but curls up and virtually disappears from sight when indoors. Another aspect of activity level that does not involve running around is vigor, which refers to the force or intensity of behaviors, rather than how fast they are. A vigorous dog will eat and drink with great gulps, snap food from its owner's hand, pull hard on the leash, dig with his paws, or push or chew with great force to get through a barrier to something he desires. A dog low on this dimension will be much gentler and less forceful in his actions.

Very High Energy Level: pointers, drovers, vermin hunters, herding dogs.

Moderately High Energy Level: fighting dogs, personal protection dogs, setters, spitz.

Moderately Low Energy Level: multipurpose sporting dogs, companion dogs, spaniels, retrievers.

Very Low Energy Level: guard dogs, sight hounds, scent hounds, draft dogs.

The first thing that you should know is that the experts really judged none of the dog breeds as being inactive. The four breeds that were judged least active of all were the basset hound, Tibetan spaniel, Pekingese, and clumber spaniel, but even these did not score in the lowest possible category for both indoor and outdoor activity. For this reason, our lowest grouping refers to only a moderately low energy level.

We would expect that herding dogs, drovers, and the vermin-hunting terriers score high on this dimension, but the great surprise to me was that sight hounds, those veritable running machines of dogdom (like the greyhound, which can run in excess of 40 miles per hour, or 65 kilometers per hour), score in the lowest group for energy level. This caused me to go back to the original ratings given by the experts. Indeed, if we only looked at outside activity levels, the sight hounds would be among the highest-rated animals. However, these dogs are also extremely inactive indoors and also rated as low in terms of their vigor, being mostly gentle dogs except when out hunting. Taken together this gives them an overall average energy level that is relatively low.

Emotional Reactivity

The next personality dimension is emotional reactivity. If we were talking about human beings, this dimension would be called *neuroticism* (if we were focusing on the high scores) or *stability* (if we were focused on the low scores). Emotional reactivity really refers to how quickly a dog's emotional state and behaviors can change. A dog that is easily excited and has mood fluctuations—from confident to cautious, from threatening to friendly, and so forth—would score high on this dimension.

Very High Emotional Reactivity: pointers, setters, multipurpose sporting dogs, herding dogs.
Moderately High Emotional Reactivity: retrievers, sight hounds, personal protection dogs, spitz.
Moderately Low Emotional Reactivity: spaniels, scent hounds, vermin hunters, drovers.
Very Low Emotional Reactivity: companion dogs, guard dogs, fighting dogs, draft dogs.

One important aspect of emotional reactivity is empathy, which is the ability to sense and share the emotions of others. It is not surprising to find setters and herding dogs in the highest group, since these breeds are often credited by people as having something almost akin to extrasensory perception when it comes to recognizing the moods and feelings of the people around them.

For the breeds that are lowest in emotional reactivity, one should not read their low scores as meaning "not emotional" but rather as "not changing quickly." This means that while a setter might be upset now but happy a minute later, a companion dog will be slower to change his mood. When he is placid it will take him longer to get angry than it might take the spaniel; however, once angry it will take him much longer to calm down.

Dominance/Territoriality

The last personality dimension I have called dominance and territoriality. It really is related to *aggressiveness* and threats of aggression on the part of the dog. An animal that scores highly on this dimension will growl to show his displeasure and back that up with his teeth if need be. He is apt to be suspicious of strangers (whether human or canine), and is not usually tolerant when disturbed, nor understanding of the intrusion of children. He is apt to guard his territory and his possessions with great force.

Very High Dominance/Territoriality: fighting dogs, personal protection dogs, guard dogs, spitz.
Moderately High Dominance/Territoriality: pointers, vermin hunters, drovers, setters.
Moderately Low Dominance/Territoriality: multipurpose sporting dogs, spaniels, herding dogs, retrievers.
Very Low Dominance/Territoriality: scent hounds, draft dogs, sight hounds, companion dogs.

In these rankings, the experts seem quite consistent with popular stereotypes. The protection, guarding, and fighting breeds all come out with the highest rankings on this aggression-related dimension. Similarly, companion dogs rank dead last, and hounds show very little aggression overall.

How Some Personality Traits Interact

It is interesting to look at the relationships between these various personality dimensions. When we statistically analyze the results of the experts' ratings, we find that the most aggressive dogs tend also to be the least sociable. Sociability and dominance are not, however, opposite ends of the same dimension, since energy level also plays a role in aggressiveness. The dogs that are most likely to show dominance and territorial aggression are both low in sociability and high in energy level. Examples are the fighting dogs, which are in the highest rank for dominance, the lowest for sociability, and second to the highest group in energy level. While one might have expected that emotional reactivity would also be related in some way to aggressive tendencies, this does not appear to be the case, according to the experts' data.

Another interesting relationship comes out of the data: a reasonable predictor of learning ability seems to be a dog's energy level. When I first observed this relationship, I was puzzled, but after a while, it began to make sense. A dog that is active, vigorous, and energetic will perform many behaviors, and will manipulate and change his environment with much more frequency than a sluggish dog will. Every time a dog does something, there is the possibility that that behavior will either be rewarded or punished. Since rewards and punishments are the basis by which behaviors are learned, the active dog simply has more opportunities to learn than the inactive dog. A low-energy dog doesn't try many alternatives and thus may not hit on the correct solution to a problem, or he may not come upon the behavior that leads to a reward simply because he works slowly and stops working early. Thus, it is not surprising that there are three breed groups (guard, sight hounds, and scent hounds) that are found in the lowest category for both energy and learning ability, while two of the most active breed groups (herding and drovers) score at the highest level of learning ability.

Unfortunately, this also means that the dog that many people dream of—namely a placid, not very active dog that is also extremely intelligent and easily trained—probably does not exist. Your brightest dogs, like border collies, are also going to be whirlwinds of energy.

Now that we have looked at the overall picture of breed differences in personality and temperament, you can hurry to the back of the book and look at the specific rankings for individual breeds and see the personality profiles for your favorite breeds. Please remember, however, that even within a breed there will be differences among individuals. For example, in one of my beginners' dog obedience classes I had two black Labrador retrievers, both of which were just under a year in age. One was active, incredibly sociable, and so busy with the world around her that she was virtually impossible for an inexperienced person to train. The other was much calmer, not as forthcoming, a wee bit shy of people but was managing to work her way through the obedience exercises. Both dogs are the same breed and even the same sex, yet had noticeable differences in their personalities. This means that knowledge of a dog's breed is not enough to give us a full and exact picture of a dog's temperament.

Think of it this way: A dog's breed tells us a lot about that dog's genetic heritage and makeup. Genetics is a strong determinant of personality. In the absence of any other information, we can make a reasonable prediction about how the dog will behave based upon its breed. For example, the typical Labrador retriever will be more sociable than the typical Akita. This does not mean, however, that if I arbitrarily select Fido the Labrador and compare his sociability to Rover the Akita, that Fido must, necessarily, be more sociable. A lot will depend upon how the dogs were reared as pups and the experiences that they have had. If Rover was well socialized and handled well as a puppy, while Fido was socially isolated and did not have close human companionship until he was sixteen weeks of age, then we might well end up with an Akita with a higher degree of sociability than a Labrador retriever.

Knowledge of the dog's breed does give us a starting place, since we know that dogs of that breed have a definite bias toward a particular personality. If, returning to our example, the Labrador and the Akita are reared in similar environments and under similar conditions, then it is a pretty good bet that the Labrador will be the more sociable. Thus, knowing the breed gives you a good starting place for predicting personality and behavior.

CHAPTER SIX

Testing a Dog's Personality

Every time you look at a dog, you spontaneously do a quick personality assessment of it. It may well be at the subconscious level of simply trying to answer the questions, "Is that dog friendly?" or "Is he one of those frightened dogs who will shy away if I try to pet him?" You are probably also doing much the same kind of assessment of the human being at the other end of the leash, as you ask yourself, "Is he in control of his dog?" or "Will he resent my coming over to say hello to his pet?" We automatically ask such questions about dogs, people, and any other living being that we encounter so we know how to react to them. We want to understand their personalities in order to predict their behaviors and adjust our own behaviors accordingly.

An entire branch of the science of psychology is devoted to developing tests and measures that can describe human personality. The resulting "personality profiles" can be used to counsel individuals to select suitable jobs or appropriate educational goals. Knowledge of one's personality can help them anticipate and cope with particular problems that they might be expected to encounter and show them the psychological resources and weaknesses that they have so that they can get along better with work mates and family members.

Not surprisingly, similar attempts have been made to systematically measure the personalities of individual dogs. Information of this sort is desirable when choosing a puppy to be a household pet, selecting a dog for a particular service occupation, or deciding whether a dog might be a successful competitor in canine sports. Measuring personality is also of great importance to people who work in animal shelters and who need to determine which dogs are adoptable, and what kind of family setting would best suit a particular dog. While adopting a dog with a bad or inappropriate temperament can be a trial for any person or family (and may lead ultimately to the dog being rejected, abandoned, or euthanized), it can also harm the reputation of the animal rescue organization that placed it with that family. If the dog turns out to be aggressive, for example, word gets out that this facility does not assess the temperament of their dogs or advise potential adopters of behavior problems. This will reduce the pool of people willing to take dogs from such a shelter and, with the loss of public support, funding will suffer. In the extreme, if a person is hurt by an aggressive adopted dog, and the shelter had given no warning, legal actions and damaging financial judgments may result.

Testing for Aggression

Aggression has become the main focus of the testing programs at animal shelters, because growls, threats, and bites are the most common behaviors that cause dogs to be returned to shelters after adoption. Because aggressive behaviors have health, safety, and legal implications, many animal shelters have begun to give temperament tests to dogs before deciding if they are adoptable. Many such tests tend to focus only on aggression.

Temperament testing has gone on informally for many years. I can remember going to an animal shelter a number of years ago to talk to the director who claimed that he used quick, efficient temperament testing procedures to determine if dogs were adoptable or needed to be euthanized.

"Come with me," he said, "and I'll show you how it's done" as he picked up a lanyard that had a pen hanging from it and looped it around his neck. He also picked up a clipboard, which had some forms on it designating dogs (by breed and kennel number) with boxes next to each, some of which had already been filled in. Finally he grabbed a long stick that looked like it had been cut from a broom handle, strode out the door of his office, and down the corridor to the entrance to kennels.

At the kennels he moved to a far row and started his "testing." In the first kennel was a mixed breed that seemed to be mostly Rottweiler in appearance. He took his stick and banged loudly twice against the metal door of the kennel. The dog leapt forward barking and growling.

"Now that's one extreme," he said, while marking something on his clipboard. "This dog is clearly aggressive and not adoptable. He is already on the terminate list. Let me show you the other extreme."

He walked down the row of kennels until he came to one that contained a somewhat smallish smooth-coated collie. Again the two loud bangs of the stick against the kennel door, and the dog started back in fear, cringing against the back wall of the kennel, and a small puddle of urine formed at its feet.

"A dog like this will be tested three times over a three-day period. If she shows signs of getting over her fear, then we'll see what we can do to make her adoptable. Otherwise . . ." He made another mark on his clipboard.

"Some people say that temperament testing is a waste of time, but as you see it doesn't take all that long. I don't have any systematic statistics, but I think that this procedure has cut down on the number of dogs returned to the shelter for aggression or fear-related problems. Furthermore, the quick decision that we can make for some dogs saves the shelter money on maintenance and prevents overcrowding."

I left the shelter in a depressed state wondering about the number of dogs that were needlessly dying because of this useless, inappropriate, and ultimately inhumane "testing procedure." It brought to my mind the fact that up until 1944 the American Kennel Club allowed judges to "test the temperament" of working dogs in the conformation ring. The procedure consisted of having the judge actually hit the dog in the face, usually with his hand but sometimes with a rolled-up newspaper or magazine. A dog that showed fear or submission when struck was judged as having a poor temperament. The kennel club and judges argued that this was a necessary test since a shy or fearful dog would not be a good guard, military, or service dog. The practice was eventually stopped after a few noisy incidents at shows where spectators protested when they observed the judges striking the dogs. A *New York Times* article on this issue argued that these protests were the actions of ignorant members of the public "who know little or nothing of the fine points of breeding or judging." However, the kennel club reacted to the generally bad reactions of observers by banning the practice. It is interesting to note that it was the dogs that showed "spirit" by barking, growling, or snapping at the judge who struck them that were deemed as having passed their "temperament test" in the ring. These very behaviors that made the dogs "good examples of their breeds" in the show ring would have put these dogs on the "terminate list" in the shelter I visited because of a similar "temperament test."

Tests That Are Useful but Not Perfect

My visit to that shelter occurred in the early 1970s and I would like to be able to report that all temperament testing in shelters now has a scientific basis and is more valid and reliable. Unfortunately, this is not necessarily the case, although some people are making great efforts to rectify this situation and there have been improvements. Most currently used temperament tests have several subtests and apply a standardized scoring system. The majority involve observing the dog's reactions to being approached, stared at, touched, manipulated, threatened, or restrained.

Some temperament test inventories set up situations where the dog is exposed to other dogs or unfamiliar people and its reactions are noted. In some cases, scenarios are set up to provoke the dog deliberately in order to see how it responds. One popular test starts by giving the dog a bowl of food. An artificial human hand on the end of a stick is then used to grab at the bowl or push it away from the dog, and its reaction (for example, biting at the hand) is supposed to measure the dog's willingness to use aggression to guard things that he feels belong to him. In other cases, a toddler-sized doll with its hands raised and extended forward is propelled at the animal to simulate a child rushing at the dog. Again, the measure is whether the dog shows any aggressiveness in its response.

Are the results of such tests meaningful? Psychologist Emily Weiss, who is now the senior director of shelter behavior programs for the American Society for the Prevention of Cruelty to Animals (ASPCA), was commissioned by the Kansas Humane Society to develop a test battery for dog aggression. The result was the SAFER test, which employs items from some of the other existing tests (like the scenarios I described above) and uses a novel scoring procedure. Weiss also conducted research to determine the validity of the tests as a measure of aggressive tendencies. In one study, she followed two groups of dogs that had been adopted from a shelter—one group had been tested for aggression and the other had not. Three weeks after the dogs had been sent to homes she phoned their new owners to do a follow-up. Starting with 141 dogs, she found that 12 were euthanized because of behavioral problems, but only 4 of these were in the group that had passed the test. Thirty-six dogs from the untested group showed aggressive behaviors in their new homes, compared to only 8 that had passed the test. In an interview she said, "We repeated the test about six months later and got similar results. After that, they [the staff at the Kansas Humane Society] were not comfortable putting dogs up for adoption that hadn't been tested."

In a similar study, Amy Marder, vice president of behavioral medicine at the ASPCA, did a telephone follow-up on 70 dogs that had been given a 140-item test focusing mostly on aggression prior to being adopted out. As with Weiss's research, Marder's results led her to conclude that the results of the tests were useful, but not 100 percent certain. Generally, if a dog showed aggression during any of the tests, it was more likely to show aggression later in its new home. However, there seems to be something about the shelter situation that can either aggravate or suppress fearful and aggressive responses. A significant percentage of dogs that responded with aggression when the artificial hand grabbed at their food bowl often calmed down and showed little or no aggression once they were living happily with a new family. However, some dogs that don't show aggression in the shelter do so later. Marder warns, "The dogs change in two directions [after they leave the shelter], an increase in behavior or decrease in behavior," which seems to be another way of saying that the predictive ability of the tests for aggression currently being given in shelters is far from perfect.

A study at Cornell University by Tracy Kroll, Katherine Houpt, and Hollis Erb, specifically assesses the accuracy of two of the most popular tests. When the artificial hand was used to grab at food or other items, approximately one-third of the dogs that had a previous history of aggression did *not* show aggression during the test. On the other hand, one-quarter of the dogs that had no history or other evidence of aggression did snap, growl, or bite during the test. The results were similar when the child-sized doll was used to provoke the dog, with approximately one-third of the dogs with no history of aggression against children (or anyone else) showed aggression during the test, while a bit less than one-quarter of the dogs with a documented history of such aggression did not snarl or snap at the doll. Thus, their results show that the tests are correct two-thirds to three-quarters of the time, which is certainly better than chance, but a long way from being truly reliable. They summarize their findings saying, "In a shelter situation, if these tests are used as a part of determining adoptability, then dogs will be missed that are aggressive, and some dogs may be deemed unadoptable that would otherwise be adoptable." However, given the potentially dire consequences of canine aggression, they suggest that any dog that shows aggression during these tests "should be looked at with concern." They offer the hope that, in conjunction with an expanded test battery with many other items and several methods of evaluation, the data from such tests will prove helpful. For the moment, however, we must conclude that tests for aggression given in animal shelters, although useful, still only give an estimate of the probability that a dog will or won't be aggressive later—they certainly don't offer certainty.

The Dog Mentality Assessment Test

People working at animal shelters are not the only ones who want to test the personality of dogs. Furthermore, by focusing only on aggression, although it is a crucial trait, we end up getting no information about other characteristics of a dog's personality that are also important. If we truly want to understand how a dog will behave in a variety of situations, we must know much more about that dog's character than simply whether he is snappish or not.

Fortunately, a number of broad-range canine temperament tests are available, some of which have some reasonable scientific data supporting their use. Most have been designed for specific purposes, since what might be an acceptable personality for a pet dog might be inappropriate for a guard dog or a search and rescue dog. The temperament tests that are probably best known and most widely used were designed in Sweden and Switzerland specifically for use by centers that select service dogs. These centers are trying to determine which animals have temperaments that are suitable for canine occupations such as police work, explosive and drug detection, search and rescue, or providing assistance to blind, handicapped, or hearing-impaired people. It is economically important to select dogs for training that have the right characteristics since it is very expensive to have a dog fail to complete a program after a lot of training time has been invested in it.

Records have been kept on thousands of the dogs tested by some European centers. Probably the largest such data bank was assembled by the Swedish Working Dogs Association, which contains data from 15,329 dogs at the time of this writing. The assessment procedure they used is called the Dog Mentality Assessment Test. One of the first uses of this particular test was to assist people who were trying to improve the breeding of working dogs. The idea was that the test would be used to compare the behavioral reactions of the puppies to that of their parents to see which personality characteristics were passed on genetically. The test appeared to be quite successful and was soon adopted by many breed clubs in Sweden, who use it as a general test of canine temperament. Variations of this test have been adopted by various other dog clubs and service dog organizations around the world. The test has also become popular with dog owners who simply want to measure their dog's personality for their own enlightenment or to see whether their dog might be suitable for certain dog sports or activities.

The Dog Mentality Assessment Test attempts to be quite objective and its procedures and scoring have been standardized so that results from one testing center can be compared validly to those from another. The process involves exposing dogs to several different, novel situations, each of which is considered to be a subtest in the battery. The dog's reactions are then evaluated by specially trained observers using a specifically designed score sheet that describes possible behaviors that might be seen in each subtest and assigns each a value. Each of the separate subtests is set up in advance, since a number of people and pieces of apparatus are required to give the entire inventory. Usually the parts of the test are arranged as a series of different stations along a path in an outdoor area. This method allows many dogs to be tested in a short period of time. For each subtest, a judge observes and scores the dog with the help of several assistants.

An Afternoon at the Testing Field

One dog organization in my area decided to set up a version of this temperament test. This group was interested in selecting dogs that would be good for both service and for schutzhund trials. *Schutzhund* is the German word for "protection dog" but has also come to be the name of a sport in which dogs engage in a number of tasks that service dogs, particularly those that work with police and military groups, often are called upon to perform. Dogs have to demonstrate three aspects of behavior in schutzhund trials. The first is tracking, which requires the dog to track a path left by a person walking over mixed terrain several hours earlier. Along the way, the dog must find various articles dropped by the person and indicate their locations to the handler. The second phase is much like standard obedience exercises, with the dogs heeling, both on and off lead, and responding to sit, down, stand, and stay commands. The dogs are also asked to demonstrate some more complex learned skills, which include controlled retrieving, jumping, and wall climbing.

The unique aspect of schutzhund trials is the third phase in which dogs demonstrate protection ability. Here the exercises include a sort of partially improvised piece of theater that enacts the process of finding and apprehending a criminal. It begins with a search of hiding places to locate a hidden person, or "decoy," who acts as the stand-in for a criminal. The dog has to guard that "criminal" while the handler approaches and then later pursue him when he attempts to escape. When the decoy is "captured," he is searched and taken to the judge with the handler and dog walking near him. When the "criminal" tries to attack the handler the dog is expected to stop the attack.

Later in the testing, another decoy rushes out of a hiding place near the judge but from the opposite end of the trial field. When that decoy "refuses" to respond to a command to stop, the dog is sent after him. Now the decoy turns and threatens the dog with a stick, and the dog is expected to attack him and hold on until he is commanded to stop or until the counterfeit criminal discontinues the fight. Obviously, this portion of the work requires the dog to have not only courage, but also a degree of aggressiveness.

Some experts from Sweden were brought in to help set up the temperament test and train the other personnel who would be needed to carry out and score it. I decided to take my flat-coated retriever, Odin, for an assessment and to see how the temperament testing procedure worked in actual practice.

The examiners prefer the dogs to be at least eighteen months old, and that no dogs younger than one year of age were accepted for testing. This is because some of the tests can be quite stressful and if the dog is too young, he might suffer some residual negative effects. Odin was just over two years of age at the time so this was no problem.

The test began with the Odin walking beside me on leash. Several people were present as audience and assistants, and there was a test leader who served as the judge. The test leader explained that I would be allowed to give my dog any help and support that he might need to overcome any fear or shyness that might be triggered by the tests.

The first test is for sociability and involves nothing more than meeting the test leader, who greeted me, then briefly ran his hands over Odin. He then took the leash from me, walked a short distance, and then returned Odin to me. Friendliness and absence of fear or aggression was all that was required for my dog to pass this test.

The second test involved playfulness and a bit of persistence. It began with the dog off leash and was designed to determine if the dog would play with a stranger. First, the "stranger" and I tossed a knotted rag pull toy back and forth to catch the dog's interest. Next, the stranger threw the toy and told Odin to fetch it. To pass, the dog has to bring the toy back to the stranger. The dog also has to be willing to play tug of war with him, but to break off from playing the game when told to. The procedure was repeated a couple of times to see if the dog quickly got bored with the game. Odin, being a retriever, did not get bored with retrieving the toy, but also did not try very hard to hold on to it during the tug of war.

In the next test, the dog was supposed to chase something. The idea behind this is to see if his predatory instincts are still present. A fuzzy rug-like piece of cloth was tied to a long string that was then threaded around posts and through pulleys, so that when the string was pulled it would move in a zigzag path over a fair distance—about 80 feet (25 meters). The person pulling the string was far enough away so that the dog couldn't see him running with the end of the line. Presumably, the bit of rug must look like a rabbit or some other form of furry prey moving quickly away. The preferred response to this situation is for the dog to run after the rug at high speed and, when he catches it, to grab and shake it. Odin did chase and grab it, but being a retriever, he carried it gently, rather than trying to shake and "kill" it.

Next came a brief interlude of calm before the storm of frightening events. This involved a test of the dog's activity level and an assessment of how it behaves when its handler was passive and not doing anything. I was simply required to stand absolutely still with my dog still on leash for a period of three minutes. I was not allowed to talk to Odin, but had to stand quietly. To pass, the dog has to be alert and calm and simply wait for something new to happen. Whining, chewing on the leash, or restless activity would fail this portion of the test. Odin waited beside me about thirty seconds and then decided that if I wasn't doing anything he would settle down for a quick nap in the warm afternoon sunshine.

In my mind, it is after this that the most interesting part of the test begins, because I was not really sure how my dog would respond. The first part involved observation of the dog's reactions to a sudden, potentially confusing or dangerous event. The test leader pointed in the direction that I was supposed to walk with my dog still on leash. There was nothing visible there but I started to walk as instructed. Suddenly a human-sized dummy, dressed in work clothes, popped up from the ground (triggered by a person pulling on a rope some distance away). My instructions had been that when this happened I was supposed to let go of the leash and let my dog react. That was easy for me to do since I had forgotten exactly where the dummy had been hidden and when it popped up I was so surprised that I dropped my leash spontaneously. Fortunately, the handler was not being scored.

This situation typically produces a variety of different responses. I later saw one dog run a long distance away in fear, another growl and bark, and yet another that had to be coaxed to draw closer to this thing that suddenly appeared. Part of the problem for the dog is that this thing looks like a human but neither acts nor smells like one. The optimal behavior in this test was to do exactly what Odin did. He acted startled for a moment, but then, on his own, moved forward to check out the situation and see if there was any danger. In a few moments, he was sniffing at the dummy with his tail wagging cautiously. I recovered the leash and walked him past the dummy several times to demonstrate that he no longer had even a trace of hesitation or fear.

The test continued as I moved on a short distance farther down the path away from the dummy. Having just recovered from an unexpected visual stimulus, Odin now encountered the startling, sudden, unexpected sound of a heavy metal chain being dragged over a sheet of corrugated metal just as we passed it. The procedure was the same as that used for the dummy. Again I had to let go of the leash and allow the dog to react. A dog with a stable temperament, like Odin's, will momentarily freeze but then move toward the source of the sound to investigate. It was interesting to watch a few other dogs at this task. Some, who had apparently not fully recovered from the dummy test, and who, according to their owners, normally did not react to loud noises, acted quite startled and unsure. The residual effects of the previous test seemed to be added to their reaction making them shy and afraid. Several balked and a few minutes later were still resisting attempts to walk them once again past the place where the noise occurred.

At the next station things heated up even more. Odin and I were walking at a normal pace when we were suddenly confronted with a man who jumped out of the bushes. He was acting the part of an assailant with his arms held high, his face in a scowl, and he gave a threatening growl. Again I dropped the leash as I had been instructed. This test is an unpredictable situation, given that many dogs become quite aroused by the two previous tests. Some dogs simply break and run in fear, some (although not very many) will leap forward and attack the man. The response that the judges are looking for is a dog that stands and barks or growls in a protective manner in front of his master. Odin stood, braced himself protectively, just in case some action was needed, but the retriever in him seemed to prevent him from uttering any sound.

Now the assailant undergoes a metamorphosis. Just as suddenly as he appeared from the bushes, he ceased to be threatening. He lowered his arms, adopted a friendly face, and stopped moving toward us. This is a test to see how quickly the dog forgives a threatening or startling social event. The desired response is for the dog to now approach the man and make some form of friendly social gesture, which Odin found quite natural to do.

The scare tactics were not over yet. We moved down the path a bit further to a place where the marker indicated that I was to stand still. Suddenly two "ghosts" appeared, one on either side of us, moving toward us. Actually, these ghosts were just assistants covered in white sheets and white hoods. The hood had big painted eyes and a vicious-looking grin that showed a lot of teeth. The sheet covered their bodies completely, so that no arms or legs were visible and they seemed to move in a flowing yet awkward manner. Once the ghosts were very close they turned around so that the frightening eyes and mouth disappeared. This test might be considered the exact opposite of the dummy test. These things smell human but don't look human. I released the leash so Odin was free to react. Later I would see one dog that had made it through the dummy and the loud sound bolt for cover, simply because the emotions aroused by these successive events ultimately built up and became too much for him. Another leapt at the nearest ghost, biting and tearing the sheet. More stable dogs stand and cautiously observe or bark protectively. Now one of the ghosts started to speak and said Odin's name while flipping up the sheet to show his arms and remove his hood. The other ghost repeated these actions a short time later. The desired behavior is that the dog approach, inspect these creatures, and make social contact once convinced that they are human. Odin approached with his tail swinging happily, almost as though he were asking to be invited to whatever costume party he thought that they must be going to.

The last test of the sequence assessed the dog's response to gunshots. Here situational factors can play a role, since a dog who is normally stable in the presence of gunshots could easily fail on a day like this where he has already encountered strange threatening people, loud noises, dummies, and ghosts. For some dogs, the emotional buildup is too great and they fail the gunshot test. Sometimes the judge or the owner recognizes the state that the dog is in and stops the test at this point, rather than potentially leave the dog in a state of fear that might have lasting effects on future encounters with gunshots. For those that continue, this last test begins with the owner and dog playing with a pull toy. Then a gun is fired from a distance of about 50 feet (15 meters). The dog is allowed to react but should immediately return to the game. If the dog is showing insecurity, or the reaction is uncertain, the gun may be fired several times to be certain of the dog's reactions. If the first gunshot results in a fearful response, no further shots are fired. Odin did look in the direction of the sounds for a moment, but then his retriever instincts swamped any other thoughts in his head and he immediately returned to me and the toy in the hopes that I would throw it for him to fetch.

The whole process took under an hour. Odin came out of the test with his tail still wagging. He was still a bit excited by everything that happened but seemed his usual happy, confident self. The judge noted that he was a very stable dog, although he had concerns about whether he had the protective instincts needed to become a good schutzhund dog since Odin had not uttered a single growl or bark throughout the entire test. I did not want to seem to minimize the value of the judge's comments, so I did not mention that Odin's not barking or growling was the feature of his performance with which I was most pleased. It appeared that all of my attempts to socialize my dog so that he was "bomb proof" and not made uncomfortable by any aspects of human behavior seemed to have paid off.

The judge's comment and my reaction to Odin's performance point out an important aspect of personality testing for dogs. Each such test is given with a specific purpose in mind. In some cases, the goal is to identify a dog that will be a good protection dog, guard dog, service dog, handicap assistance dog, hospital visitation or therapy dog, or perhaps just a quiet calm companion dog. There are no "right" or "wrong" answers to a personality test. "Passing" or "failing" a personality test is really just a matter of determining whether the profile of traits in the dog being tested is a good match to the profile of personality traits in dogs that have been successful at particular tasks, or in particular environments. A dog that is too aggressive to be a good family dog might well be the perfect guard dog, while a pleasant, unconcerned dog that would be a great family pet would have exactly the wrong personality characteristics to be a sentry patrol dog.

Interpreting the Personality Tests

Kenth Svartberg, at Stockholm University, and Bjorn Forkman, at the Royal Veterinary and Agricultural University in Frederiksberg, Denmark, gathered data from more than fifteen thousand dogs that had taken the Dog Mentality Assessment Test (just like the test that Odin took) and subjected the results to a rigorous scientific analysis to see what it said about canine personality. As I mentioned in the previous chapter, they had found that the personality structure of dogs seemed to be described by five basic personality traits. While the number was the same, the traits that they uncovered were not exactly the same as the five traits that Gosling was looking for or the five traits that my experts came up with; however, there were similarities.

One of the characteristics of data like this is that at the best of times, there is a lot of overlap between the various traits. This is one reason why different researchers sometimes decide upon different labels for the personality characteristics that their research reveals. Recognizing this, Svartberg and Forkman tried to simplify things by doing further statistical analysis and found that they could combine all of the traits, except aggressiveness, to form a broad personality characteristic called the "shyness-boldness continuum." Dogs that are high on this personality trait are bold, usually very active, interested in other dogs and people, curious and relatively fearless when faced with novel objects and strange situations. Dogs that score low for this trait are shy, tend to be uninterested in play, and are timid, cautious, and evasive in unfamiliar situations. Other research has shown that this shyness-boldness dimension is also found in wolves, although as a group, wolves are much more likely to be found on the shy end of the continuum than dogs.

Svartberg was particularly interested in finding a way to use a dog's personality characteristics to predict his performance on tasks related to military and police work. Such tasks might include not only tracking or searching, but might require the dog to protect his handler. In a separate piece of research, Svartberg concentrated on German shepherds and Belgian Tervurens to see if an individual dog's personality characteristics, specifically its standing on the shyness-boldness continuum, would predict later success as service dogs. One of his more interesting results confirmed a popular stereotype that males are generally bolder than the females. However, when it comes to carrying aggression to the point of actually biting, females were more likely to carry through, although the difference was small. The results also showed a breed difference, with the German shepherds being the bolder of the two breeds. The main finding, however, was that the dogs with the highest boldness scores, regardless of their sex or breed, tended to do the best when trained for service work. Bold dogs seem easier to train as working dogs. Bold dogs are also more active, which confirms the findings of my experts who showed that dogs that are more active are more trainable and appear to be more intelligent and better problem solvers.

This analysis also seems to confirm the usefulness of this personality test as a means of predicting behaviors in dogs. However, here we must be careful. Many people want a way to test the personality of puppies in order to predict their behaviors as adults and thus find them appropriate homes and work settings. Let's remember, however, that the Dog Mentality Assessment Test used here has an age restriction. The test requires the dogs to be at least eighteen months old, which is far beyond puppyhood. Are there personality tests that can be successfully used on puppies?

Testing Puppies

Obviously, the earlier that dogs with appropriate temperaments can be selected for certain service jobs, the better things will go in the long run. Fewer man-hours will be wasted and less money will need to be spent training dogs that will ultimately prove incapable of doing the required work. For this reason, many attempts have been made to design tests that can validly measure the personality of puppies. For example, Clarence Pfaffenberger, one of the most important figures in the development of training and selection programs for guide dogs for blind people, used a variety of tests to select dogs for this task. He claimed that a young puppy's willingness to retrieve playfully thrown objects was the best single indicator of whether it would grow up to be a good working dog and used this as one of the criteria in selecting guide dogs.

Large-scale, systematic attempts to evaluate the personality of puppies and predict their behaviors as adults were first attempted in Europe. One of the earliest was the Fortunate Fields Project in Switzerland. During the 1920s and 1930s, their goal was to develop the perfect service dog. The only dog breed tested in this project was the German shepherd, but the testing system and scoring procedures include personality tests that can be used for any breed. The Fortunate Fields' methods of testing would ultimately evolve into the Dog Mentality Assessment Test that I described earlier.

William Campbell, a well-known animal psychologist and one of the founders of the American Society of Veterinary Ethology, tried to create a test that could be used for assessing puppies. He wanted it to be simple enough so that it could be used by breeders and dog owners at home. Knowing of the work done by Pfaffenberger and the Fortunate Fields Project, he used their findings as a jumping-off point. Campbell looked at four personality traits. The first was *excitability versus inhibition,* where a more excitable dog is very responsive to any form of stimulation while an inhibited dog is much more self-controlled. This is much like the trait of emotional reactivity or neuroticism. The second was *active versus passive defense tendencies,* where an active response to a threat is biting while a passive one is less confrontational and involves freezing or running away, which appears to be much like Svartberg's boldness trait. The remaining two dimensions indicated whether the dog was *dominant versus submissive* in social settings, which is virtually identical to the dominance/territoriality dimension found in my experts' data, and a *sociability* factor that measured whether a dog sought to make contact with people, or was independent and more of a loner, which is the equivalent of the sociability factor found in every test that we have looked at so far.

Since Campbell's initial work, several other tests have looked at the personality of puppies. Perhaps the most popular of these was developed by the talented and innovative dog behaviorists Joachim and Wendy Volhard. Their Puppy Aptitude Test took elements from many earlier tests (Fortunate Fields, Pfaffenberger, and Campbell's tests), and added two tests suggested by Elliot Humphrey and Lucien Warner for working dogs, one for *sound sensitivity* (a sound-sensitive dog becomes fearful around loud sounds such as gunshots or shouted commands) and *touch sensitivity* (a touch-sensitive dog can be difficult to train since any collar or leash correction can set off the dog's defensive reactions). Together these last two tests probably reflect the emotional reactivity or neuroticism trait. The Volhards integrated these into one system with an easily interpretable scoring system. Although some dog behaviorists have modified the testing procedures or scoring for some special purposes, such as fitness for obedience training, the Puppy Aptitude Test remains the most popular personality testing procedure in North America and is used by many breeders and prospective dog buyers to evaluate puppies before choosing one from the litter.

Unfortunately, some recent scientific studies have questioned the usefulness of puppy temperament testing. One of the largest, best-controlled studies to determine whether puppy temperament tests predict adult behavior was conducted by Erik Wilsson, of the Department of Zoology at Stockholm University, and Per-Erik Sundgren, of the Department of Animal Breeding and Genetics at the Swedish University of Agricultural Sciences. They used records from a set of objective behavioral tests originally developed by the Swedish Army School and which were later standardized by the Swedish Dog Training Center. Specifically the pups were tested for sociability, independence, fearfulness, competitiveness, general activity, and exploration behavior. The test battery also included a retrieving test similar to that used by both Pfaffenberger and the Volhards. After they were tested, the puppies were placed in private homes for socialization and basic obedience training and were later brought back to the center for further testing between the ages of fifteen and twenty months. Again, only one breed was used, so breed differences in temperament did not influence the findings. Data was collected from 630 German shepherd pups, and personality test scores at eight weeks were compared to those at approximately a year and a half. The results were disappointing and showed no evidence that the puppy scores predict the adult personality. Several other studies using smaller samples have arrived at much the same conclusion.

Now some of you might feel that I am contradicting what I have said earlier, when I noted that personality tests have been shown to be useful in predicting the success of service dogs. Let's examine another large study by Wilsson and Sundgren, since these researchers went on to test 1,310 German shepherd dogs and 797 Labrador retrievers on ten personality dimensions and found that the test predictions were sufficiently reliable to allow the researchers to determine which combinations of personality traits would help the dogs to succeed (or fail) when assigned to learn a specific job.

What determines the difference between successful predictions and unsuccessful predictions? The important factor is the age of the dogs when they were tested. In the second, more successful study, the dogs were tested in their adolescence or young adulthood, namely between fifteen and twenty months of age. One reason the age of testing is important is that the dogs were now out of their puppy stage, when so many changes occur so quickly. For example, how a dog is socialized, how much experience it gets with people, other dogs, different rearing environments, and so forth, will continue to change and mold the dog's temperament well beyond the age of eight weeks when puppies are usually tested. It appears that by the age of around a year and a half a dog's personality is much more likely to have been permanently fixed. With this later testing, the results are quite different and much more useful and reliable. Thus, personality tests are not valid when given to puppies, but these tests, when given to adolescent or young adult dogs not only work to determine which dogs will become good service dogs, but also seem capable of determining which dogs would be most successful as human companions.

Fortunately, a few important personality dimensions are established early enough to be reliably measured in puppies. For instance, fearfulness is a vital personality dimension, since it is the most common reason why dogs are disqualified from guide dog programs. Michael Goddard and Rolf Beilharz, at James Cook University in Australia, studied the trait of fearfulness in dogs. Tests given to pups at the age of twelve weeks reliably predicted fearfulness in the adults. However, the predictions were even more reliable if they were taken at six months of age, much like testing for other aspects of personality. This means that, although we can detect an enduring tendency to be fearful in early puppy testing, testing an older dog gives a better picture of what the adult personality characteristics will look like.

This whole discussion of testing puppies reminds me of a friend who has had a number of dogs with good stable temperaments that have gone on to do quite well in obedience competitions. I once asked her, "How do you usually test the personality of the puppies that you are thinking about getting?"

She smiled and explained, "I lift each puppy up, hold his face near me, and look it in the eye. The first puppy that doesn't act fearful or annoyed and licks my face is usually the one that I go home with."

I suspect that the good temperaments of her dogs has more to do with how she rears and trains them once she has taken them home than with the accuracy of her testing procedures. On the other hand, it is rather pleasant looking closely into the face of a puppy. Who knows? It might work as well as some of the other tests currently in use.

CHAPTER SEVEN

The Dog Behavior Inventory

In recent years, the search for a simple way to measure the personality or temperament of dogs has taken an interesting turn. James Serpell, of the Veterinary School at the University of Pennsylvania, and Yuying Hsu, of the Department of Biology at National Taiwan Normal University, reasoned that there should be a way to modify the questionnaire methods typically used to gather opinions on the personalities of dog breeds from experts, so that an individual dog owner could determine the personality profile of his own dog. Their research begins with the presumption that no one knows more about a dog's typical behavior than a person who lives with that dog. They reasoned that it should be possible, by asking appropriate questions, to extract accurate and reliable information about the dog's personality from its primary caretaker.

This resulted in the development of a questionnaire modeled after those that are used by psychologists to get information about the behaviors and personality of young children. What dogs and young children have in common is that they can't fill in a personality test and are not verbal enough to explain their feelings and behaviors. Instead, psychologists rely on information provided by the child's parents or caretakers. Typically, the parent is given a questionnaire or checklist and asked to report how their child would react to certain situations or how often that child would show certain behaviors. For every item, the parent judges how strong or frequent that behavior is by checking off a scale that has five to seven alternatives. The answers are then converted into numerical values to allow statistical analysis to determine the personality profile of that particular child.

Serpell and Hsu modified this procedure to determine personalities of dogs. Their first study evaluated the temperaments of dogs that were candidates to become guide dogs for the blind. The information was provided by the volunteer dog raisers who brought the pups into their homes, cared for them, and lived in close association with them until the dogs were ready to start training. The researchers found that the answers provided by the people described the personalities of the dogs accurately and reliably enough to predict whether their dogs would successfully complete the guide dog training program.

Given the success of this study, they attempted to develop a broader questionnaire to assess a wider range of personality traits (not just those that were expected to predict the success of service dogs). The original version of what is now called the Canine Behavioral Assessment & Research Questionnaire (C-BARQ) contained 152 items, although it has currently been reduced to 101 items. The questions varied in formats, from estimates of how frequently certain behaviors appeared, to rating scales estimating the intensity of certain behaviors or behavioral predispositions. Thus, one question might present a situation and ask the caretaker to estimate whether their dog typically showed fearfulness under such circumstances. The responses available would be five alternatives ranging from "no fear" to "extreme fear." Another question might ask whether in a particular setting the dog would typically be described as "calm" up to "extremely excitable" with several intermediate levels between. Statistical analyses of the responses from several thousand people showed that the questionnaire data was reliable and could accurately predict a number of different personality traits in dogs. The accuracy of the predictions was verified by the fact that these personality profiles could be used to screen dogs as to the likelihood that they would develop certain behavioral problems.

The Serpell and Hsu questionnaire is a fine research instrument, but its size and use of multiple-question formats and scales of different lengths makes it a bit cumbersome for the average person to use at home to assess the personality of his own dog, or dogs. Furthermore, the particular traits it measures are somewhat different from those that we have been discussing thus far in this book. Nonetheless, this research demonstrates that the technique of asking those living with a dog to use their "expert knowledge" by observing their own dog's behaviors does produce a potentially useful picture of a dog's personality.

Let's look at a simpler questionnaire to assess a dog's personality, one that I have been using in my laboratory. Psychologists who read this will notice that I have borrowed a trick from Sam Gosling (who did the research on the similarity between canine and human behavior) in that a number of the questions in this survey are based upon items that appear in human personality tests, only these have been modified to capture canine rather than human behavioral tendencies. At the time of this writing, I have collected data designed to assess the personality of 617 dogs, based upon their owners' answers to this questionnaire. This number of animals should provide a large enough data sample to allow us to determine statistically how common certain behavior patterns are among dogs in general and to allow you to compare your own dog's personality traits to those of other dogs.

I called my questionnaire the *Dog Behavior Inventory* (or DBI). It is quite straightforward to use. We start with the assumption that your dog is at least one year of age (a year and a half or more in age will probably produce more accurate results) and that he has been living with you long enough so that you have a good feeling for how he typically acts. Each item in the questionnaire describes a dog's behavior and perhaps a situation where your dog might show such behaviors. You simply have to consider your own dog and determine how typical that behavior is for him. You will have five choices for each answer, *Never* (which really means never or almost never), *Seldom, Occasionally, Frequently,* or *Always* (or almost always). Simply check the box to the left of your answer. Try to be as honest and accurate as possible. If you are in doubt, it may be worthwhile to check with other family members to see what their observations have been. Try to answer every question, since the scoring falls apart if any items are left blank.

Dog Behavior Inventory (DBI)

This first set of questions deals with the ***energy level*** of your dog

1. Is your dog playful, puppyish, or boisterous?
☐ Never[1] ☐ Seldom[2] ☐ Occasionally[3] ☐ Frequently[4] ☐ Always[5]

2. Is your dog active, energetic, and always on the go?
☐ Never[1] ☐ Seldom[2] ☐ Occasionally[3] ☐ Frequently[4] ☐ Always[5]

3. Does your dog wake up quickly and seem ready for anything immediately?
☐ Never[1] ☐ Seldom[2] ☐ Occasionally[3] ☐ Frequently[4] ☐ Always[5]

4. Do you find that it is hard to go fast enough, play hard enough, or stay out long enough to satisfy your dog when out for a walk or exercise?
☐ Never[1] ☐ Seldom[2] ☐ Occasionally[3] ☐ Frequently[4] ☐ Always[5]

5. Does your dog like being involved in games or being the center of attention?
☐ Never[1] ☐ Seldom[2] ☐ Occasionally[3] ☐ Frequently[4] ☐ Always[5]

6. Does your dog knock things over or bump against people or things in its haste to do things?
☐ Never[1] ☐ Seldom[2] ☐ Occasionally[3] ☐ Frequently[4] ☐ Always[5]

7. Does your dog tend to quickly and vigorously snap food or treats from your hand when they are offered?
☐ Never[1] ☐ Seldom[2] ☐ Occasionally[3] ☐ Frequently[4] ☐ Always[5]

(8) Does your dog like to lie quietly even when people or other dogs are noisy and active nearby?
☐ Never[5] ☐ Seldom[4] ☐ Occasionally[3] ☐ Frequently[2] ☐ Always[1]

(9) Is your dog slow and deliberate in most of its actions?
☐ Never[5] ☐ Seldom[4] ☐ Occasionally[3] ☐ Frequently[2] ☐ Always[1]

(10) Does your dog do more resting than moving?

☐ Never[5] ☐ Seldom[4] ☐ Occasionally[3] ☐ Frequently[2] ☐ Always[1]

The next set of questions has to do with your dog's ***learning, obedience, and problem solving.***

11. Does your dog obey simple commands like "sit" and "down" immediately?

☐ Never[1] ☐ Seldom[2] ☐ Occasionally[3] ☐ Frequently[4] ☐ Always[5]

12. When your dog is off leash and running free, does it return immediately when you call?

☐ Never[1] ☐ Seldom[2] ☐ Occasionally[3] ☐ Frequently[4] ☐ Always[5]

13. Does your dog retrieve or try to retrieve sticks, balls, or other objects if you throw them for him or her?

☐ Never[1] ☐ Seldom[2] ☐ Occasionally[3] ☐ Frequently[4] ☐ Always[5]

14. Does your dog seem to pay attention or listen closely to everything you say or do?

☐ Never[1] ☐ Seldom[2] ☐ Occasionally[3] ☐ Frequently[4] ☐ Always[5]

15. Does your dog figure things out for itself (e.g., he or she may be good at escaping from places, learning how to open closets or doors, quick at understanding words like "walk" or "bath" that you haven't deliberately tried to teach him, et cetera)?

☐ Never[1] ☐ Seldom[2] ☐ Occasionally[3] ☐ Frequently[4] ☐ Always[5]

(16) Is your dog easily distracted by interesting sights, sounds, or smells?

☐ Never[5] ☐ Seldom[4] ☐ Occasionally[3] ☐ Frequently[2] ☐ Always[1]

(17) Is your dog slow to learn new tricks or tasks?

☐ Never[5] ☐ Seldom[4] ☐ Occasionally[3] ☐ Frequently[2] ☐ Always[1]

(18) Does your dog give up quickly when it can't solve a problem?

☐ Never[5] ☐ Seldom[4] ☐ Occasionally[3] ☐ Frequently[2] ☐ Always[1]

(19) Does your dog repeat the same mistakes over and over when having difficulty solving a problem or learning something new?

☐ Never[5] ☐ Seldom[4] ☐ Occasionally[3] ☐ Frequently[2] ☐ Always[1]

(20) Is it the case that your dog knows what to do but can be stubborn and refuse to obey commands?

☐ Never[5] ☐ Seldom[4] ☐ Occasionally[3] ☐ Frequently[2] ☐ Always[1]

The next set of questions has to do with ***how your dog interacts with your family and others.***

21. Does your dog greet everyone as if they were a friend?

☐ Never[1] ☐ Seldom[2] ☐ Occasionally[3] ☐ Frequently[4] ☐ Always[5]

22. Does your dog come to strangers if they call or offer their hand or a treat?

☐ Never[1] ☐ Seldom[2] ☐ Occasionally[3] ☐ Frequently[4] ☐ Always[5]

23. Does your dog tend to follow you (or other members of your household) around your home as you go from room to room?

☐ Never[1] ☐ Seldom[2] ☐ Occasionally[3] ☐ Frequently[4] ☐ Always[5]

24. Does your dog tend to stay close to, or in physical contact with, you (or other members of your household) when you are sitting down?

☐ Never[1] ☐ Seldom[2] ☐ Occasionally[3] ☐ Frequently[4] ☐ Always[5]

25. Does your dog like to play or interact with children?

☐ Never[1] ☐ Seldom[2] ☐ Occasionally[3] ☐ Frequently[4] ☐ Always[5]

26. Does your dog like to play or interact with other dogs?

☐ Never[1] ☐ Seldom[2] ☐ Occasionally[3] ☐ Frequently[4] ☐ Always[5]

(27) Does your dog act distant, reserved, or aloof around strangers?

☐ Never[5] ☐ Seldom[4] ☐ Occasionally[3] ☐ Frequently[2] ☐ Always[1]

(28) Does your dog act like a "one-family" or "one-person" dog?

☐ Never[5] ☐ Seldom[4] ☐ Occasionally[3] ☐ Frequently[2] ☐ Always[1]

(29) Does your dog leave the area or move away when others are actively playing or socializing?

☐ Never[5] ☐ Seldom[4] ☐ Occasionally[3] ☐ Frequently[2] ☐ Always[1]

(30) Does your dog like to spend time by himself (or herself)?

☐ Never[5] ☐ Seldom[4] ☐ Occasionally[3] ☐ Frequently[2] ☐ Always[1]

The next section has to do with your dog's ***emotional reactions*** in common situations.

31. Is your dog easily startled by loud noises or unexpected events?

☐ Never[1] ☐ Seldom[2] ☐ Occasionally[3] ☐ Frequently[4] ☐ Always[5]

32. Is it common for your dog to quickly change moods (for example, going from being comfortable in a situation to being wary or suspicious)?

☐ Never[1] ☐ Seldom[2] ☐ Occasionally[3] ☐ Frequently[4] ☐ Always[5]

33. Can you accurately predict how your dog will react to a new or strange situation or when meeting new people or dogs?

☐ Never[1] ☐ Seldom[2] ☐ Occasionally[3] ☐ Frequently[4] ☐ Always[5]

34. Is it easy to work your dog up to a high pitch of activity when you or others are playing with him?

☐ Never[1] ☐ Seldom[2] ☐ Occasionally[3] ☐ Frequently[4] ☐ Always[5]

35. Does your dog act nervous or high-strung?

☐ Never[1] ☐ Seldom[2] ☐ Occasionally[3] ☐ Frequently[4] ☐ Always[5]

36. Do changes in routine seem to bother your dog?

☐ Never[1] ☐ Seldom[2] ☐ Occasionally[3] ☐ Frequently[4] ☐ Always[5]

37. When people are loud or angry nearby, does your dog react?

☐ Never[1] ☐ Seldom[2] ☐ Occasionally[3] ☐ Frequently[4] ☐ Always[5]

(38) Is your dog generally in good spirits?

☐ Never[5] ☐ Seldom[4] ☐ Occasionally[3] ☐ Frequently[2] ☐ Always[1]

(39) Does your dog seem comfortable in new and unfamiliar places?

☐ Never[5] ☐ Seldom[4] ☐ Occasionally[3] ☐ Frequently[2] ☐ Always[1]

(40) Is your dog calm and easygoing?

☐ Never[5] ☐ Seldom[4] ☐ Occasionally[3] ☐ Frequently[2] ☐ Always[1]

The next set of items looks at your dog's ***boldness, territoriality, dominance, and assertiveness.***

41. Does your dog bark, growl, or bare his teeth when mailmen or other delivery workers approach your home?

☐ Never[1] ☐ Seldom[2] ☐ Occasionally[3] ☐ Frequently[4] ☐ Always[5]

42. Does your dog assume a dominant position around other dogs (e.g., standing tall and stiff with tail up, staring directly at them, or trying to mount them)?

☐ Never[1] ☐ Seldom[2] ☐ Occasionally[3] ☐ Frequently[4] ☐ Always[5]

43. Does your dog bark, growl, bare his teeth, or snap when verbally corrected or punished, scolded, or shouted at by you or another household member?

☐ Never[1] ☐ Seldom[2] ☐ Occasionally[3] ☐ Frequently[4] ☐ Always[5]

44. Does your dog bark, growl, bare his teeth, or snap when approached directly by an unfamiliar adult or a child while being walked or exercised on a leash?

☐ Never[1] ☐ Seldom[2] ☐ Occasionally[3] ☐ Frequently[4] ☐ Always[5]

45. Does your dog bark, growl, bare his teeth, or snap when toys, bones, food, or other objects are taken away by a household member?

☐ Never[1] ☐ Seldom[2] ☐ Occasionally[3] ☐ Frequently[4] ☐ Always[5]

46. Does your dog bark, growl, bare his teeth, or snap when approached or disturbed by a family member, child, or another dog while it is in a favorite resting or sleeping place?

☐ Never[1] ☐ Seldom[2] ☐ Occasionally[3] ☐ Frequently[4] ☐ Always[5]

47. Does your dog bark, growl, bare his teeth, or snap when being bathed, clipped, or groomed?
☐ Never[1] ☐ Seldom[2] ☐ Occasionally[3] ☐ Frequently[4] ☐ Always[5]

48. When your dog is in the yard, on the porch, at the window, or near the door, does he bark, growl, or bare his teeth when joggers, cyclists, Rollerbladers, skateboarders, or people walking dogs pass your home?
☐ Never[1] ☐ Seldom[2] ☐ Occasionally[3] ☐ Frequently[4] ☐ Always[5]

49. Is your dog insistent in his demands for food, play, affection, et cetera (this includes the dog's nudging, nuzzling, pawing at you, barking, whimpering, or staring directly at you to try to get your attention and to get you to respond)?
☐ Never[1] ☐ Seldom[2] ☐ Occasionally[3] ☐ Frequently[4] ☐ Always[5]

(50) Does your dog seem to ignore or sleep through events that you believe would excite other dogs?
☐ Never[5] ☐ Seldom[4] ☐ Occasionally[3] ☐ Frequently[2] ☐ Always[1]

This last section deals with your dog's ***anxieties and fearfulness.***

51. When you are away from your home or away from familiar surroundings, does your dog crouch or cringe (with tail lowered or tucked between the legs), whimper, whine, freeze, tremble, or actively try to avoid or hide when approached directly by an unfamiliar adult?
☐ Never[1] ☐ Seldom[2] ☐ Occasionally[3] ☐ Frequently[4] ☐ Always[5]

52. When you are away from your home or away from familiar surroundings, does your dog crouch or cringe (with tail lowered or tucked between the legs), whimper, whine, freeze, tremble, or actively try to avoid or hide when approached directly by an unfamiliar child?
☐ Never[1] ☐ Seldom[2] ☐ Occasionally[3] ☐ Frequently[4] ☐ Always[5]

53. Does your dog crouch or cringe (with tail lowered or tucked between the legs), whimper, whine, freeze, tremble, or actively try to avoid or

hide when it encounters strange or unfamiliar objects on or near the sidewalk (e.g., garden or cleaning tools and equipment, plastic trash bags, litter, flags flapping, gates or doors swinging, et cetera)?

☐ Never[1] ☐ Seldom[2] ☐ Occasionally[3] ☐ Frequently[4] ☐ Always[5]

54. Does your dog crouch or cringe (with tail lowered or tucked between the legs), whimper, whine, freeze, tremble, or actively try to avoid or hide in response to sudden or loud noises (e.g., thunder, vacuum cleaners, car backfire, road drills, objects being dropped, et cetera)?

☐ Never[1] ☐ Seldom[2] ☐ Occasionally[3] ☐ Frequently[4] ☐ Always[5]

55. Does your dog crouch or cringe (with tail lowered or tucked between the legs) or roll on its back when approached by an unfamiliar dog?

☐ Never[1] ☐ Seldom[2] ☐ Occasionally[3] ☐ Frequently[4] ☐ Always[5]

56. Does your dog tend to panic or frighten easily?

☐ Never[1] ☐ Seldom[2] ☐ Occasionally[3] ☐ Frequently[4] ☐ Always[5]

57. Does your dog crouch or cringe (with tail lowered or tucked between the legs), whimper, whine, freeze, tremble, or actively try to avoid or hide when first exposed to unfamiliar situations (e.g., first car trip, first time in elevator, first visit to veterinarian, et cetera)?

☐ Never[1] ☐ Seldom[2] ☐ Occasionally[3] ☐ Frequently[4] ☐ Always[5]

58. Does your dog dribble urine when approached by an unfamiliar person or dog, or when brought into an unfamiliar place?

☐ Never[1] ☐ Seldom[2] ☐ Occasionally[3] ☐ Frequently[4] ☐ Always[5]

59. If someone, some place, or some thing frightens your dog, does your dog seem to remember and respond fearfully the next few times that he or she has to deal with it?

☐ Never[1] ☐ Seldom[2] ☐ Occasionally[3] ☐ Frequently[4] ☐ Always[5]

60. Does your dog seem hesitant and seem to need to check things out carefully before acting?

☐ Never[1] ☐ Seldom[2] ☐ Occasionally[3] ☐ Frequently[4] ☐ Always[5]

After you have answered all of these items for your dog, you next have to determine your dog's scores, which you will enter in the table below. First, note that the general scoring system is that *Never=1, Seldom=2, Occasionally=3, Frequently=4,* and *Always=5.* However, for the questions that have parentheses around their numbers, the scoring is reversed going from *Never=5* down to *Always=1.* To simplify matters I've included the score for each answer in the form of a superscript number next to each. Simply add the score for the ten questions in each group and fill in the table below. I'll tell you what to do with the rating column afterward.

Dog Behavior Inventory Scores

Personality Trait	Sum of Questions	Total	Rating
Energy	1–10	______	______
Intelligence/Learning ability	11–20	______	______
Sociability	21–30	______	______
Emotional Reactivity	31–40	______	______
Dominance/Aggressiveness	41–50	______	______
Fearfulness/Anxiety	51–60	______	______

Now we have to interpret these scores for your dog. This interpretation will be based on data that I obtained from 617 dog owners who reported on the characteristics of their own pets using this questionnaire. The sample of dogs was quite mixed: 38 percent (234) were crossbreeds of various sorts, while the remaining 62 percent were purebred dogs (383) representing all of the various dog groups. We'll use a rating method for your dog's personality traits that is consistent with the ratings by dog experts that I discussed in chapter 5. These are given in detail in the appendix so you'll be able to see whether your own dog differs from a typical member of its breed.

The ratings represent a relative measure of how a dog ranks in comparison to all of the dogs that were tested in my data set. We rank the dogs based upon dividing them up into quartiles (25 percentile groupings), just as we did for the experts' rankings of the personality traits of the 133 breeds listed in the appendix. Thus ***very low*** is the lowest 25 percent compared to other dogs ranked for each trait (75 percent of all dogs score higher than this); ***moderately low*** is higher than the lowest 25 percent of the dogs but lower than 50 percent of the dogs ranked for each trait; ***moderately high*** is higher than 50 percent of the dogs but lower than the top 25 percent of dogs ranked for each trait; ***very high*** is the highest 25 percent compared to other dogs ranked for this trait (75 percent of all dogs score lower than this).

First, let's look at *energy.* In general, this composite measure looks at the dog's indoor and outdoor activity level. It also includes a measure of vigor, or the amount of force and energy the dog will bring to common activities. Thus, a dog high on energy is always in motion and its actions are vigorous (meaning that it may tug strongly on the leash, and so forth), while a dog low on the energy dimension will be much quieter, more inactive most of the time and gentler in its activities. In human personality systems, this is most closely related to the dimension *extroversion* where active extroverts are contrasted to the more passive and retiring introverts. The average score for this trait is 34. Dogs with scores less than 27 are very low energy, scores from 27 to 34 are moderately low energy, above 34 up to 41 would rank the dog as moderately high on energy, while scores above 41 make a dog high in energy. Simply mark your dog's ranking in the space beside its score.

The second personality trait measured in this inventory is *intelligence and learning ability.* This is a measure of how easily a dog learns and solves problems. It is also an indication of how easy it will be to train the dog to do various things, ranging from simple dog obedience commands through to more complex tasks, such as those required of service dogs. In humans the aspect of personality associated with this trait is *openness,* and it is not only associated with intelligence but also with creativity and imagination. People high in this trait also tend to be more daring, with broader interests and much more curiosity. Individuals lower on this tend to avoid new experiences and prefer more familiar predictable situations. The average score for this trait is 30. A dog with scores below 23 is very low, between 23 and 30 is moderately low, from 30 to 37 is moderately high, and above 37 is very high for this trait. Write your dog's rank in the space provided.

The third trait measured by the DBI is *sociability,* which refers to how friendly a dog is, and how much it seeks out companionship. A dog high on this trait will happily greet any new person, while a dog low on this trait may appear shy and aloof, may take a long time to warm up to new acquaintances, and may prefer to be alone rather than with people most of the time. In human personality terms, this is usually labeled *agreeableness,* and is a reflection of whether the individual is warm and pleasant or cold and distant. Generally speaking, individuals who are high in this dimension also tend to be cooperative and trusting. Dogs high on this trait are easy to work with but may demand a lot of attention. The average score for this trait is 31. Dogs with scores below 24 are very low in sociability, while dogs with scores between 24 and 31 are moderately low. Scores from 31 to 38 are moderately high, and scores above 38 are very high in sociability.

The fourth trait is *emotional reactivity,* a measure of how quickly the dog's mood may change and how his emotional state fluctuates. Dogs that are high on this dimension will be easy to excite but will also calm down easily. They will also startle easily and owners will generally have more difficulty predicting exactly how their dog will react in any new situation. A dog that is low on this trait will be more difficult to excite but will also take much longer to calm down. In humans, this trait is called *neuroticism* and is usually contrasted with its opposite pole, *stability.* In humans, this trait is also associated with emotional tone. People high in this trait are more apt to be moody, anxious, insecure, and tend to worry constantly for no discernable reason. People low on this trait are much calmer, have fewer major mood swings and are usually described as "easygoing" or "unflappable." On the whole, individuals who are low in emotional reactivity tend to be more satisfied with their lives, performance, and conditions. The average score for emotional reactivity is 29, with scores below 22 being very low, those between 22 and 29 ranked at moderately low, those from 29 to 36 being moderately high, and those above 36 ranked very high.

The fifth trait measured by the DBI is *dominance* and *aggressiveness,* which also includes how *territorial* the dog is. Dogs high on this trait tend to be bold, self-assured, confident, and pushy. Such dogs make good watchdogs and guard dogs, but they also may be quite vigorous and possessive when guarding their own things, such as food and toys. Dogs high on this personality trait are quite willing to use threats or, if necessary, their teeth, to assert their status or defend that which they consider to belong to them. These are characteristics that often make a dog good at certain service occupations, but can make them difficult (or at the highest levels, dangerous) in the average family and neighborhood settings. Although we can see similar traits in humans, dominance and aggressiveness are not considered primary traits associated with most human personality theories. In dogs, however, probably because of their evolutionary role as predators, it is a fundamental characteristic of their temperament. It is obviously also important in determining how well a dog will fit into human society. The average on this scale is a bit lower than for other scales at 26. Dogs with scores below 19 are very low in dominance and aggression, between 19 and 26 dogs rank moderately low, while dogs between 26 and 33 are moderately high on this trait, with the dogs scoring above 33 showing very high levels of this characteristic.

The final characteristic measured here is *fearfulness* and *anxiety,* which pertains to how easily the dog is frightened, how nervous and insecure the dog appears to be, and how frequently the dog tries to run away from or avoid situations because they arouse negative emotions in him. Most people mistakenly think that a frightened dog is not going to be an aggressive threat. This is false. A frightened dog may appear timid, hand-shy, or just plain "spooky" with a desire to stay away from some people and situations, but if its level of anxiety is high enough, and the option of escaping from the fear arousing situation is cut off, such dogs can become aggressive. Fearfulness and anxiety are the principal reasons dogs flunk out of service dog training programs. You simply cannot rely on a guide dog for the blind, for instance, who is apt to panic in just the sort of situation where its master most needs its guidance. The average score for fearfulness and anxiety is 28, with dogs below 21 being very low, dogs between 21 and 28 being moderately low, dogs between 28 and 35 being moderately high, and dogs above 35 being very high in fearfulness.

In all of these rankings, especially if you are looking at the moderately high or low rankings, it is important to be aware of the average for the group. For instance, with fearfulness, the average score for the group is 28. If you have a dog that scores 29, although it would be classified as moderately high, it is really pretty much an average dog. For all of these scales, dogs that score within two or three points of the average should be considered fairly typical dogs—that is, not really very high or low, but right in the middle of the group for that trait.

It would be nice to be able to go through all possible combinations of personality ratings so that with any dog's pattern of scores we could determine what to expect of that dog's behavior. Unfortunately, given the fact that we have six dimensions of personality and four possible scores on each, the total number of combinations of traits (think of these as personality types) is 4,096. It would take a book that is substantially larger than this one to describe each separately (although you might think of dropping a note to my editor if you want me to produce such a book). Nonetheless, I can give you some general guidelines for the combinations of traits that make certain types of dogs good for particular settings or tasks.

All-around family dog: low to very low energy. Any intelligence rating can work reasonably well except very low. High to very high sociability is preferred along with low to very low for emotional reactivity, dominance, and fearfulness. An active and athletic family or someone who walks a lot and spends much time outdoors might want higher energy rankings.

Quiet companion dog: low to very low energy. Intelligence can be at any level if someone is at home most of the time (however, if the dog must spend considerable time alone the extreme scores of very low and very high intelligence can lead to problems). Avoid very low sociability. Emotional reactivity, dominance, and fearfulness should all be low to very low.

Obedience competition dog: high to very high energy, intelligence and sociability are the keys here, combined with low to very low emotional reactivity and fearfulness. It is probably best to avoid dogs that are very high in dominance.

Competitive sports dog (agility, fly ball, et cetera): high to very high energy and intelligence, sociability is more flexible, although very low should be avoided. Emotional reactivity should be in the mid range (low or high are acceptable but not very low or very high). Avoid very low dominance, but low to very low fearfulness is best.

Service dog (sentry dog, patrol dog, schutzhund, et cetera): high to very high energy and intelligence. Avoid very high sociability. Emotional reactivity should be in the mid range (low or high are acceptable but not very low or very high). Dominance should be high or very high. Very low fearfulness is preferred, although you can get away with low fearfulness.

Service dog (search and rescue): only a few differences from the protection or guarding service dog. Look for high to very high energy and intelligence; avoid very low sociability. Emotional reactivity should be in the mid range (low or high are acceptable but not very low or very high). Dominance can be mid range (low or high) and can be very high if sociability is high or very high. Very low fearfulness is preferred although you can make do with low fearfulness.

Assistance dog (guide for blind, handicap assistance dog): Very high energy dogs should be avoided. High to very high intelligence is required, and sociability should be high to very high. Emotional reactivity and dominance should both be in the mid range (avoid very high or very low). Very low fearfulness is best, although low is sometimes adequate.

Hearing assistance dog: A few differences from the other types of assistance dog; very-low energy dogs should be avoided. High to very high intelligence, sociability, and emotional reactivity are preferred. Dominance can be anything other than very low, and fearfulness should be very low, although low is often adequate.

Therapy and hospital visitation dog: Energy level should be low to very low, intelligence level is not that important for hospital visitation dogs although for some therapy dogs with much interaction very low intelligence is best avoided. Sociability should be high or very high, emotional reactivity should be low or very low, dominance is best close to average (can be low or high), while very high dominance must be avoided and very low is acceptable if sociability is very high. Fearfulness should be low, although very low is preferable.

I suppose that after indicating some useful combinations of traits, it might be worthwhile to look at the worst possible combinations.

The Dog from Hell

Generally speaking, a combination of very high emotional reactivity, very high dominance, and high to very high fearfulness is bad. It may sound contradictory to suggest that you can get both high dominance and high fearfulness in the same dog, but remember that the dominance scale reflects not merely the dog's feelings about his social status, but also his territoriality and possessiveness, but also how likely he is to use aggression. Thus, the very high dominance score can reflect a dog that uses aggressive behaviors willingly, and those behaviors can be fueled by high fearfulness and strong emotional swings. Combine this with very low sociability and high to very high energy levels and you are on your way toward creating the ultimate devil dog. Strangely, intelligence contributes only at the extremes. Given very high intelligence these dogs can become nightmares as they figure out how to do many forbidden things and how their actions provoke desirable reactions from the humans in their environment. Given very low intelligence, this combination of other traits turns the dog into the canine equivalent of the unrestrained comic book character, the Incredible Hulk. Such dogs react with raw physical actions driven by overwhelming and unpredictable emotions. These devil dogs are often the result of poor early socialization or bad genetic programming; they are often impossible to live with and sometimes a danger to family members and others that they might meet.

Although the research of Serpell and Hsu demonstrated that questionnaires, like the DBI, could give us a good picture of a dog's personality, there is a weakness in this method. It depends upon the honesty of the dog owner when answering it, and many owners are simply blind to any bad behavior on the part of their dog. For example, one woman who had taken our beginner's dog obedience class and was currently enrolled in our advanced beginner's class, had a Jack Russell terrier that always ran forward and snapped and growled at the nearest dog when it entered the door. Her response was always an amazed "I don't know why Zorro did that. He's never done that before. He's always good with other dogs." I'm sure if she filled in the DBI questionnaire, Zorro would score a lot lower on the dominance and aggression scale than he would if I filled out one based on my observations of him.

CHAPTER EIGHT

Creating a Superdog

Although up to now we have emphasized the genetic contributions to personality, we have already encountered some evidence that it is possible to shape personality in ways other than selective breeding. For example, had you seen my big black flat-coated retriever, Odin, during the first few months that he lived with me, you would have clearly thought that I had brought a monster into my house. The genes of the flat-coated retriever are programmed to produce a bouncy, sociable, intelligent, playful dog that loves to retrieve and run in the field. That produces a puppy that comes into the house and bounces off furniture and walls, jumps up and deposits muddy paw prints on anyone who enters a room, and fails to respond to any human command when he is busy chasing something that might be retrievable. To turn him into the still friendly but calm, "bomb-proof" dog that was not disturbed, angered, or frightened by anything humans did near him, did not involve genetic engineering, but rather giving him an appropriate set of early experiences.

We can influence the behavioral predispositions of a maturing dog because its personality, like that of a human, is not fully formed at birth. Furthermore, a puppy is born with such an immature, incomplete nervous system that, by controlling its experiences we can even shape the structure of its brain. This gives us an opportunity to exert a huge influence on how he'll behave as an adult.

For example, the brain of a day-old puppy that might eventually grow up to be a 65-pound (30-kilogram) dog is extremely small. It is only about 10 cubic centimeters, the size of an average adult's middle finger from the tip to the first joint. Obviously, that brain still has a lot of growing to do. By the time the pup reaches eight weeks of age, his brain will have increased to more than five times its original size and will be close to 60 cubic centimeters. After another eight weeks of growth, it will increase another 25 percent and should be around 80 cubic centimeters. Not until sometime between the ages of nine months and one year does a midsized dog's brain reach its full volume of about 100 cubic centimeters. This means that the day-old pup starts out missing 90 percent of the brain matter that it will have as an adult.

The new puppy's brain is not only very small but it is still incredibly immature in its structure. It appears to be almost jelly-like, since the fibers that connect the neurons have not yet developed the fatty white sheath (called myelin) that speeds communications between locations in the brain and electrically insulates each nerve cell so that its messages are not interfered with by the action of other neighboring cells. The neural responses of the brain are also quite primitive in the newborn pup. In fact, an EEG (a measure of the electrical activity of the brain) shows responses that are so rudimentary that a trained scientist finds it difficult to even tell whether the puppy is awake or sleeping. When a puppy is ten days of age, we can easily count the number of neural connections (synapses) that a single cell in the cortex has with other cells in the brain since there will be only a few hundred. By the time the puppy reaches thirty-five days of age, however, the number of connections for each neuron in the brain will have multiplied to around 12,000.

Measurements of the newborn puppy's brain—including size, weight, the amount of oxygen that it consumes, and pattern of electrical activity—all suggest that for the first six or seven weeks of life the pup's brain is growing at an astonishing pace. That rate of development will begin to slow after week seven, although substantial growth will continue until the dog reaches around six to eight months of age.

Environmental Influence Can Start Early

Although the puppy has many behavioral characteristics that are coded in his genes and are predictable because of his breed, his psychological destiny, his temperament and personality, and even his final intelligence level are not predetermined and immutable. Many of a dog's most important behaviors are *epigenetic*, which literally means "above the genes." The puppy's experiences and interactions with the environment, especially those that occur at an early age, will have a great deal of influence on the adult dog's behavior. I am not speaking only of what the puppy will learn from rewards or punishments for specific behaviors. Some other special forms of learning that seem quite different from any found in humans will influence the later behavior of the puppy.

Our understanding of how to create a dog with a better and more stable personality begins, strangely enough, with some cold rat pups and some laboratories in the biosensor division of the United States Army. First, it is important to recognize the long history of research that has tried to identify how early experiences and life history shape the behaviors and personalities of adults. The psychologist Sigmund Freud loved to emphasize how important early experiences were with a quote from the poet William Wordsworth, namely, "The child is father to the man."

Many people think that environmental influences begin at birth, but scientists recognize that environmental influences actually begin virtually from the moment an egg is fertilized. A dog's brain and behavior are shaped to some degree by events in the womb. Research has shown that when a pregnant dog is stressed, there are negative effects on the pups. Most typically, if a mother is stressed, while pregnant, her offspring will be timid and grow up to be relatively fearful animals. As in much of psychological development, timing is extremely important. The most vulnerable period seems to be when the mother is in the final third of her pregnancy. Severe stress at this stage will produce pups with greatly reduced learning ability. Pups that have grown in the womb of such a stressed mother may show some extreme or exaggerated behaviors, and their responses to new conditions and people may range from freezing in fear to the opposite extreme of being exceptionally emotionally reactive and trying to escape from the situation. Researchers believe that this is a direct result of chemical events in the womb. Stress causes the release of a number of hormones, particularly corticosteroids, and changes in the later behavior of the pups are a direct result of the effects these chemicals have on the brains of the unborn puppies.

An interesting additional finding is that the puppies, while still in the womb, are also being influenced by their littermates depending on the sex of the other pups in the litter. The sexual characteristics of an animal are determined by the presence or absence of certain hormones. The basic design of all mammals is female, and only the presence of certain male hormones (androgens) causes the individual to develop the body and behaviors of a male. Research has shown that an animal that is genetically programmed to be a male will look and act like a female if it is not exposed to these male hormones. The flip side of the coin is also true and an animal that has been genetically programmed to be a female will look and act like a male if she is exposed to high enough levels of androgens when she is still developing in the uterus. It is through this mechanism that littermates can influence each other.

Sex hormones produced by each pup in the womb seep out of the fetuses and are mixed into the amniotic fluid that surrounds all of the pups. Thus, the amount of androgens in the uterus is determined by the number of male pups. Brain and other neural tissues seem to be particularly sensitive to the presence of male hormones like testosterone. If a litter is mostly males and there is a large amount of male hormones in the fluids surrounding the developing puppies, this will affect the brains and behaviors of all of the pups developing in that womb. Females born into mostly male litters tend to act more masculine. This means that they are more active and sometimes more dominant and territorial. They occasionally may even be seen lifting their leg when they urinate, which is a characteristically male behavior. The males born into predominantly male litters then become even more pronounced in their male behaviors. For some breeds, dogs born in mostly male litters may be extremely pushy and controlling. Having the increased levels of androgens is not necessarily a bad thing, especially when you are looking for service or patrol dogs with a high level of drive. Obviously, pups born into predominantly female litters will show somewhat reduced levels of typically male behaviors in comparison.

The Handling Effect

Once the pups are born, humans can begin to manipulate their environment and personal histories so that we can create dogs with better personalities. Puppies' sensory systems are not fully functioning at birth. They are born with their eyes closed so they are effectively blind, and their ear canals are closed rendering them effectively deaf. However, the pup can taste and smell and is also sensitive to touch, pressure, movement, temperature changes, and pain. These sources of stimulation are vital, however, since, for the first week or so the puppy's behavior is mostly driven by outside stimulation. The mother must lick the pups to provide the touch stimulation to start the process of elimination and encourage digestion. They even need these sensations to feed properly, since newborn pups will only attempt to nurse when they are stimulated by their mother.

Temperature changes seem to be particularly important to the pup. As long as he keeps against something soft and warm (usually its mother and littermates) he will lie quietly for hours, but a drop in body temperature seems to be stressful. If the pup is taken away from the warmth of its mother and littermates and placed on the floor a short distance away, it begins to crawl slowly, throwing its head from side to side apparently searching for its mother's scent and a heat source. As it moves, it whimpers or yelps and apparently feels real distress. Evolution has provided an additional source of sensory information to help the puppy at this critical time in the form of special heat sensors in his nose. Yngve Zotterman, of the Swedish Research Council, discovered these unusual receptors, which are sensitive to heat that is radiated from warm objects in the form of infrared energy, and are located around the nostril slits and the openings to the nasal passages of puppies. When the puppy swings his head and detects the heat source that is his mother, he follows this "heat trail" back to her. When he touches her body, he stops whimpering and curls up against her.

Once relaxed, he drifts off to sleep. Mature dogs do not seem to have these special heat-detecting receptors, so we must conclude that they disappear or become nonfunctional as other senses mature.

A bit of folklore suggests that newborn puppies are incapable of learning at this stage of their life and probably won't benefit from interactions with their environment until they are two or three weeks of age. This is so widely believed by people who work with very young puppies that it is even repeated in books and veterinary manuals. The research shows, however, that this is not true. Although their sensory systems are immature and sight and hearing are not functioning, these little animals can still learn and show adaptive behavior. For example, they can learn smells. In one study, researchers daubed the nipples of a nursing mother with anise oil during the first five days of feeding. The pups that had been fed by her then learned that that scent signified their mother. If you then took a cotton swab that had been soaked in anise and placed it near these pups they would turn toward it and wriggle their way closer. Pups that had not had early exposure to this smell found the scent to be too pungent for their tastes and would actually turn away from the anise-scented swab and try to escape from it.

Similarly, young pups can learn the significance of particular touch sensations. If, for example, puppies are placed on a textured surface (either soft or rough) and then fed from a bottle, when given the choice between a soft or rough surface they will prefer to crawl their way onto the surface that they have been fed on before. This all means that the puppies are, in fact, capable of learning from their interactions with their environment, virtually from birth, as long as the information that they are receiving is coming through those sensory systems that are already working and capable of picking up the relevant details that provide meaningful information.

Considering the helplessness of the pups, one would assume that they need to be protected from any disturbance or stimulation that might stress them. Such an assumption is all the more likely since we know that stressing the mother during her pregnancy produces animals that are deficient in learning and overly emotional. Therefore, the temptation is to keep the pups warm and quiet, avoid bothering them, and let them sleep. However, some research also suggests that the rules change once the puppies have left the womb so that leaving them in blissful peace is not the optimal course of action. To understand why, we will have to look at those cold rat pups that I mentioned earlier.

Many mammals, including cats, mice, and rats, are born with the same limitations as a newborn puppy. They too have restricted use of their senses and are easily stressed by events such as cold or high levels of stimulation. Their nervous systems are similar to that of a dog (although their brain structure is less complex) and the effects of certain environmental events seem to be the same as those observed in dogs. This means that data obtained by studying these animals can teach us what to expect in the maturing dog as well.

The first observations of importance came about completely by chance. In a number of research laboratories rats are bred to be test subjects for various behavioral and physiological experiments. When rat pups are born, they are tagged or marked with an indelible marker so that each one can be identified. The litter of rat pups is then monitored to make sure that they are developing normally and receiving adequate nutrition. Optimally, this would involve at least weighing and inspecting each rat pup daily. Unfortunately, funding for many labs is not lavish, so sometimes shortcuts need to be undertaken to cut costs. One way to save lab technicians' time, and hence money, is to select two rat pups from each litter and weigh only these daily. It is assumed that they will serve as an indication of the health of the whole litter. If one or both animals do not seem to be thriving, then the entire litter can be checked more carefully and treated appropriately. The process of removing a rat pup from its litter, placing it on a scale to weigh it, then measuring its length, noting its apparent health status, and recording the results, means that each animal is away from the nest for around three minutes or so.

Over time, technicians and researchers began to notice that the animals that had been selected for measurement and monitoring were behaviorally different. They were less fearful and emotional than their littermates, apparently learned better, and seemed more capable of coping with stressors in their environment. In some ways, they even appeared to be more physically sound. Since this occurred with such regularity, it was clear that it was not a matter of chance where technicians had accidentally picked out the most psychologically and physically sound animals to monitor. At first these findings were simply noted with some curiosity and casually labeled "the handling effect," since the rats that received this early additional handling by lab personnel seemed different. Later, however, several laboratories began to study these effects more seriously and systematically.

The Benefits of Early Stress

What was actually happening during the handling process? Apparently removing the animals from their cages placed the young pups under a certain degree of physical and psychological stress. One major physical stress had to do with temperature changes. As each animal was removed from its nest for weighing and measuring, its body temperature began to fall slightly, especially when it was placed on the cold metal surface of a typical laboratory scale. In addition, they were being physically stimulated by the technician's hand as he lifted them, moved them, and tilted them during the process of measuring, weighing, and carrying the animals from place to place. Although to the technician, all of this seemed fairly innocuous, to rat pups these were stressful events. The stress level was mild, of course, but sufficient to stimulate the sympathetic nervous system, which arouses us so that we are ready to run from danger or stand and fight for our lives. The sympathetic nervous system causes the release of stress-related hormones from the adrenal and pituitary glands, and possibly other sources.

Psychologists have clearly established that high levels of stress, especially if repeated often, are harmful. However, exposure to low levels of stress at an early age actually seems to have beneficial effects. When tested later as adults, the animals that received regular but small doses of stress in their infancy seemed to be inoculated to later stress in their lives. They were better able to withstand new and even severe stressors, with less emotional response and reduced levels of physical arousal, than their littermates who were not exposed to this early stimulation.

The way in which the early-stimulated animals responded to stress once they matured was also different. As adults, those animals that had been stressed as pups responded to stress in "a graded" fashion, while their non-stressed littermates responded in an "all-or-nothing manner." What this means is that the early stressed animals would react mildly to a mild stress and more vigorously to a strong stress. In contrast, those that were not stressed when young responded almost as strongly to mild stressors as to severe ones. Apparently, the additional stimulation early in life acts much like an inoculation. Through inoculations, individuals are exposed to a weak infectious agent that prevents major infections later on. Here we are not looking at diseases, but rather, reactions to stress.

The stronger reactions of the rats that were not stressed when young can have devastating effects. Their severe emotional and sympathetic nervous system responses can easily exhaust the body's and nervous system's resources if exposure to stressors is prolonged. This overreaction can cause damage to vital organs, including the heart and gastric system, which can eventually lead to death. For example, a major stressor for a rat is to be restrained so that it cannot move or run away. This causes an intense and primal fear response that involves all of the body's defensive systems. When rats were immobilized in this way and unable to move for twenty-four hours, the animals that had been protected, and did not have early stressors, developed severe stomach ulcers. Their littermates who had experienced mild stress as pups were found to be much more resistant to stress and showed no evidence of ulcers or other major damage.

Although we are primarily interested in behavior, it is interesting to note that at the physical level, the early stressed animals were found to be more resilient. They seemed to mature more quickly and even reached sexual maturity at an earlier age. Their immune systems were stronger and they were able to resist infectious diseases (and even certain forms of cancer) better than their unstressed littermates. When placed under severe, life-threatening conditions of cold or starvation, they had higher survival rates as well.

As researchers further explored the effects of early stress stimulation, some used a reversed method of testing. They reasoned that if stress and stimulation levels higher than those normally obtained while pups were in the litter were beneficial, then reducing the stimulation levels below the amount normally provided by mothers and littermates could be harmful. The data soon proved them correct and showed that the more that individuals were deprived of stimulation, such as social and physical interactions during early development, the less they seemed able to cope with stressful situations as adults. Furthermore, their abilities to adjust and later adapt their behaviors to meet the requirements of new and unusual conditions seemed to be greatly impaired. When such isolation and protection from early stimulation was tried on very complex animals, such as monkeys, the isolated individuals grew up to be overemotional and psychologically impaired adults.

Some of the effects of early stimulation and stress clearly caused changes in brain responses. On an electroencephalogram (EEG), which measures not only brain activity but also emotional reactions, animals that received early stimulation showed mature patterns of complex brain responses earlier. When complex problem-solving and learning tasks were set up for them, those animals were less likely to be swamped by emotional responses and stress. In addition, they seem to be effectively more intelligent, since they solved the problems more quickly and learned to apply these solutions to successive problems more efficiently than their unstressed counterparts did.

There are certain important restrictions on the early stimulation that we should note, however. First, if the early stress is too strong or too prolonged, harmful effects do occur. To have a positive impact on adult behavior, the stressors that the animal receives must be mild, relatively brief, and occur when the animals are quite young. Researchers have still not determined the optimal amount of stress and stimulation, since there may well be differences between species of animals and perhaps even breeds of dogs; however, we have a pretty good idea of the minimum amount needed to have a noticeable positive effect.

The Army's Superdog Program

This brings us to the United States Army's contribution to our knowledge of dogs, which occurred when the military attempted to apply these findings in order to rear the "perfect dog." The army's motives were quite practical, since the use of dogs in the military is widespread. Although the exact number of dogs employed across all of the military services is classified, we can get some idea from information which was released by the 341st Training Squadron, Lackland Air Force Base, San Antonio, Texas, which operates the Defense Military Working Dog Program for the U.S. Air Force. More than 125 military personnel are working full-time as dog trainers in this facility, which contains 62 training areas and spreads over 3,350 acres. There are 691 kennel spaces in this facility alone, and trained dogs are shipped out to military installations around the world. Given that the air force is only one of the services with dog training facilities, it is likely that thousands of dogs are probably being trained for military purposes each year.

Although many dogs are trained for patrol use only, which includes scouting, searching, and attacking, others are trained for property protection. One of the most important uses of dogs is in detection work, which involves finding explosives, firearms, and drugs. Dogs are also called upon to locate land mines. In various countries that have been subjected to war and civil conflict, such as Afghanistan, Bosnia, Mozambique, Nicaragua, and Rwanda, two or more people, on average, fall victim to land mines each hour. The military have more than seven hundred dogs assigned to locating mines in these countries at the time of this writing. The dogs are faster and more efficient than mechanical and electronic detection means, which is especially important given that there are millions of acres of land contaminated with mines to be dealt with.

Back in the 1960s the U.S. Army became interested in using dogs for detecting explosives, including land mines. What prompted this was the fact that manufacturers of mines had begun to use plastics and other nonmetallic components that were invisible to the metal detectors normally used to find mines. In addition, magnetic mine detectors do not work well in areas where metal content is high in the soil, or where there are many bits of metal scattered on the surface or below, such as in former battle locations which are apt to be strewn with expended shells, bullets, and bits of metallic debris. None of these conditions is a problem for dogs, however, because they are sensing the explosive chemicals directly; the presence or absence of metals is irrelevant to them.

The Army Veterinary Corps started a Biosensor Research Division to do research on the mine detection problem. The idea was to have animals (mostly dogs), which would be trained to locate mines and other substances of interest. The term "biosensor" indicates that animals, rather than electronic sensors, were being used to find explosives. The division aimed to develop techniques that would produce dogs that were physically fit and had the sound temperament and high intelligence needed to perform these services. Financial, as well as efficiency considerations, were behind this effort. To train and maintain such a mine detection dog can be very expensive. The process takes, on average, about twenty months, and can cost in excess of $100,000. Most of this cost is simply the salaries paid to the personnel and the handlers who are training the dog over this extended period. Obviously, with such an investment of time and money in each dog it is important that the dog be physically sound, so that it can have a reasonable lifespan and a long, successful time in active service. For example, German shepherd dogs were known to develop hip dysplasia, a crippling genetically related disease that could end a dog's usefulness in the military. One success of this program was to develop breeding programs to eliminate hip dysplasia from the potential army dogs.

Having a physically sound dog is not enough, however. The dog has to have an appropriate and stable personality. It is a fact that dogs are more likely to be dropped from service dog training programs because of psychological and behavioral problems than for physical shortcomings. Traveling to war-torn countries can be stressful, and if the dog does not react well to stress, its ability to perform is impaired. If the dog is not sociable around humans and does not take instruction well, it will be difficult to train. If the dog is spontaneously aggressive, it will complicate relations with civilians and others in the work area and may be a potential danger to the people it encounters. For all of these reasons, an applied research program was begun with the goal of creating a dog with a sound personality.

It has proven difficult to get full access to the records of the research from the biosensor program. Some of the documents describing the research are still considered classified, but through the U.S. Freedom of Information Act some material was made available to me. From this, I learned that the operation was set up at Edgewood Arsenal, some sixty miles north of Washington, DC, near Baltimore, under the command of Colonel M. W. Castleberry. The most complete information available deals with the group of German shepherd dogs bred between 1968 and 1976 and subjected to various special early stimulation conditions. Detailed records were available for 575 animals representing four years of testing. The animals were born under controlled conditions and came from a set of eighteen sires and seventy-one dams. At this facility, in addition to an unspecified number of researchers and veterinarians, there were also forty-eight human handlers involved in the training and testing of the dogs. A few reports that made their way out to the popular press referred to this research as the "Superdog Program."

The biosensor rearing program attempted to improve the personality of the dogs by providing the pups with mild stress early in their life. Colonel Castleberry described what the researchers were trying to develop this way:

"Temperament is the big problem. You want a bold, outgoing, self-confident dog. I define temperament as that ability in a dog to induce respect from a human being. We don't want dogs that are fearful, slinky, apathetic, that snap at you when you're not looking or that are out-and-out vicious. A puppy that runs right out and greets you in a friendly manner—that's a damn good puppy."

"The chief problem of temperament is in developing a dog that is calm and cool, but not one that is so calm and cool that he doesn't care."

Michael W. Fox, who was then with Washington University in St. Louis and ultimately would go on to become the scientific director of the Humane Society of the United States, was one of the researchers who helped set up the stimulation procedures used by the army and was also called upon to assess their effects. Puppies bred as part of the program were subjected to a full hour of intense controlled stimulation for four to five weeks. They were exposed to flashing lights and a variety of sounds. They were exposed to cold stress by putting the pups into a refrigerator for a minute each day, they were put in a device that looks like a tilted merry-go-round for a short period and then on a sort of teeter-totter device that tilted them back and forth. They were physically manipulated, including touching of their paws, mouths, and other body parts. They were put in a bath that was deep enough to require them to swim, and then rubbed dry and brushed. From three weeks of age on, they had a daily play period with their human researcher. Some of these experiences were clearly more than mildly stressful for the pups, because, as one researcher described it "There was a lot of unhappy yelping and squealing especially when they were cold-stressed, tilted, and whirled, when they were very young."

The results of these experiments proved that puppies that were handled and stressed in this manner grew up to be more confident, less fearful, and were better problem solvers. They also explored their world more and seemed less likely to be emotionally upset by unexpected events, loud sounds, and bright lights later in life. They seemed more socially dominant compared to pups that had not had such early experiences. Overall, they tended to learn more quickly and to remember better the things that they had learned. Indeed, they did seem to be the "superdogs" that the army was seeking.

Unfortunately, all was not perfect. Some pups found the stimulation far too intense, and eventually failed some of the temperament tests. The fact that there are individual differences among dogs, and a suggestion that the basic program of stimulation might have exceeded the optimal levels of early stimulation, was later verified by Fox using beagles instead of the army's German shepherds.

The best research to date has greatly toned down the intensity and duration of stimulation, and evidence is that a milder form of early stress produces all of the benefits of the Superdog program, without triggering negative effects or damaging the personalities of even the least resilient puppies. Overall, this modified Superdog program produces smarter and more stable pups than you would get if you simply left the pup alone at this time (as some books on dog rearing advise), or just presumed that its mother and littermates will provide adequate stimulation. Active handling of the puppy by humans is all that is required, and one does not need any special equipment. Let me outline how such a program of early stimulation should go.

The Early Stimulation Program

This should be started two days after the puppy is born and continued until the pup is around four weeks of age. Take each pup in turn and go through the entire sequence before going on to the next puppy.

- Motion stimulation
    - Hold the puppy in both hands with its head higher than its tail for about 10 seconds.
    - Next change the pup's position so that its head is lower than the tail for another 10 seconds.
    - Repeat this gentle slow rocking once more with head up for 10 seconds and head down again for another 10.
- Thermal stimulation
    - Hold an ice cube in your closed fist for about 10 seconds to cool down your hand, and then slip the cold hand, palm up, under the pup. He may wiggle a bit, but since your hand will quickly warm to body temperature, you are really only providing a mild stress for a short period.
    - An alternate way to do the same thing is to place the pup on a cool surface, such as a wet towel that has been put into the freezer for 30 seconds and removed, and to leave him on the towel for 10 seconds each day. (Do not put the puppy in the refrigerator or freezer as was done in the original study.)
- Tactile Stimulation
    - Hold the pup on its back and cradle it for a minute while you gently stroke its belly, head, mouth, and ears with your fingers.
    - Finally, take a cotton swab and gently spread the pads of the feet and tickle the pup between the toes.

This series of activities should take about three to five minutes at the most. Some pups may resist these exercises while others seem to take them in stride with no complaints. No matter how the puppy responds, you should give the exercises only once a day and not repeat them until the next day. You also should not extend the length of time or the number of repetitions for each. Remember, overstimulation of the nervous system can have detrimental effects, and we are trying to keep the stimulus level in the mild-to-moderate range. The idea is to "jump-start" the nervous system into action earlier than would happen without our assistance. This produces an improved mental response capacity that will make a difference in the later performance of the maturing dog. It produces all of the positive emotional effects of the more intense stimulation, the pups mature faster and appear to be brighter, and if done correctly there is no risk of the procedure overwhelming the puppy's young mind.

Although I have just warned you about overdoing these exercises, it is important to remember that this early-stimulation program is not a substitute for routine handling and socialization for the pups. Playing with them, engaging in social interactions, and other activities should—and must—be going on at the same time if we are going to produce the superdog that we desire.

Michael Fox used a somewhat toned-down version of the biosensor stimulation procedure on beagles and compared them to littermates that had only had the usual amount of stimulation that puppies generally experience. When put in a new environment the stimulated pups were far more active and explored the region to a much greater degree than their unstimulated littermates. In comparison, the dogs that had not had the early stimulation seemed more emotionally swamped and explored quite timidly if at all. Socially, the stimulated pups were more dominant over the unstimulated pups in competitive situations. In a learning test in which the pups had to solve a series of detour problems, the unstimulated puppies again ran into emotional problems and seemed much more excited and aroused. They yelped and whined and were clearly more emotionally disturbed than their stimulated littermates. Their emotional responses resulted in their making many more errors in the learning tasks. Fox summarized the comparative performance of the early stimulated animals by noting that they "kept their cool, making few errors and solving the problems very quickly and with little distress vocalization."

A Fast Start Is Not Enough

It would be nice to say that at this point the early stimulation experience was all that was needed to create our superdogs. Unfortunately, this is merely the start of the process and more work has to be done to ensure that perfect personality.

The U.S. Army program had two goals. The first goal was to create fine physical specimens of dogs that would be healthy, free of genetic defects, and have a good life expectancy. The second was to create dogs with high intelligence and appropriate personalities and behavioral characteristics to become good service dogs. Since the program was being run by army veterinarians, it is not surprising that they placed major emphasis on the physical aspects of development. Psychologists and other behavioral experts, such as Fox, were brought in to carry out tests and procedures that might influence the later personality and behaviors of the dogs. These behavioral specialists, however, worked in the program only for a limited amount of time. When the psychologists left to pursue other research, the veterinarians involved in the project returned to their major interests, which involved the physical, rather than the behavioral, status of the dogs. Some of the superdogs were placed in training and service programs and proved to be extremely successful military dogs. Unfortunately, many of the best dogs were kept in the lab to be used in later selective breeding studies. The early stimulation program continued with subsequent litters of pups, since this provided definite physical benefits in the form of improved cardiovascular performance in terms of both heart rate and strength of heartbeats, stronger adrenal glands, more tolerance to the physical effects of stress, and greater resistance to disease. Beyond this early experience, however, the dogs that should have been superdogs were simply kept in kennels and virtually ignored.

In the late 1970s the biosensor dog research program was terminated and the dogs that remained in the testing facilities were transferred to the Lackland facility that belonged to the air force. Carmen L. Battaglia, a dog expert and director of the American Kennel Club, described the condition and fate of the superdogs. "The dogs resembled a German shepherd, but most had a faraway look in their eyes. The initial litters had been raised in a kennel environment. When the dogs were exposed to common elements of an outside environment (birds, grass, et cetera), they were afraid, confused, and skittish." They were so timid that they would refuse to cross a brightly painted line on the street. They had simply been socially isolated and understimulated for too long.

Later attempts were made to train these dogs for patrol or sentry work. The dogs that were trainable did exhibit the superior intelligence that the behavioral researchers had found. However, most were unable to complete the training because of emotional problems associated with fearfulness or high emotional arousal levels that made them difficult to control. In the end, many of the dogs who could not make it through the programs were ultimately destroyed.

Does this indicate that the military program failed and we cannot really create the superdogs we are seeking? Not at all. The early-stimulation program was designed to give these dogs a good start—the equivalent of getting the runner out of the starting blocks quickly. Unfortunately, once the behavioral scientists ceased their research, no one continued the development of the dogs so that they would become better working dogs. It was as if they had trained an athlete to start a race quickly and then simply left him to figure out how to run the rest of the race on his own.

That was the problem with the superdogs that eventually arrived at the Lackland facilities. These dogs had been given a head start in their personality and mental functions, but those early advantages were lost by not continuing after the basic twelve weeks of the biosensor program. By restricting the experiences of these dogs through confinement to their kennels and providing limited continuing stimulation, these dogs that had been endowed with strong physical and neural equipment because of their breeding and early stimulation eventually became emotional cripples. If we want superdogs, we must continue beyond the first weeks in life and teach them how to run the rest of the race.

The potential in these superdogs is clearly shown by one named Bruno, who received the early-stimulation program but was not kept for later breeding since he was a bit undersized. Rather, he was given to a law enforcement agency that trained him to do drug detection work. Bruno's career included the detection of in excess of $110 million worth of smuggled drugs, including one container of narcotics hidden in a truck beneath some forty crates of oranges, and another stash of drugs in a plastic container that had been submerged in a truck's gas tank. Although not trained in protection work, Bruno's intelligence and fearless personality also helped him save the life of his handler when a drug smuggler's accomplice attempted to sneak up on the officer from behind with a knife. Bruno thwarted the ambush by barking and knocking the assailant down. He was a prime example of that bold, self-confident, resourceful, and intelligent dog that Colonel Castleberry had in mind when he started the biosensor program.

There is an odd postscript to the biosensor program. Despite its early successes, much of the data would apparently be ignored in later service dog creation programs in the United States. However, other nations were paying attention to these results. During the 1970s when the superdog program was most active, members of the Australian military consulted with the biosensor researchers and began their own breeding and rearing programs. Ultimately, the Australian Customs Service used these biosensor program techniques as the basis for starting a breeding and rearing program to produce its own version of the superdog. One major difference between the U.S. Army biosensor program and the Australian Drug Detection Dog Breeding Program is that the Australians not only worked on the genetics of their lines of dog and carefully controlled their first few weeks of rearing and early stimulation, but also made sure that the dogs would have continuing stimulation for the next year of their life. Matters thus came full circle in 1998, when the Australian Customs Service not only donated foundation breeding stock to the U.S. Customs Department to enable it to enhance its drug detector dogs program, but also gave it rearing instructions to create better dogs—the early portion of these rearing instructions are mostly based on the virtually lost and forgotten findings of the U.S. biosensor program. In effect, the Australians have demonstrated that if you follow through after the early-stimulation program, you can create superdogs after all.

Let's now see how we can go from the fast start of the early stimulation to win the whole race with our superdogs.

CHAPTER NINE

Socialization and the Superdog

If you look at the notebook of any canine behavioral consultant, you will see two large classes of difficulties that people have with their dogs. The first has to do with fear-based problems. We can list these as something like "Lassie is———" where the blank is filled in with "shy," "timid," "skittish," "easily frightened," "afraid of men with hats (or beards or glasses)," "frightened by the sound of trucks (or vacuums, thunder or crowds)," "uneasy around children (or large dogs or flickering lights)," and so forth. The second group of problems have to do with aggression, and run something like "Rover is———," where the blank is filled in with "dominant," "pushy," "snappish," "suspicious of men in raincoats (or dark-skinned men or men smoking)," "aggressive toward children (or puppies or other dogs)," "hard on the fingers when he takes treats," "barks violently at everyone who enters the house (or is seen across the street)," "doesn't tolerate being touched (or approached or even looked at)," "tends to frighten people with his jumping (or snarling or staring) behaviors." While at first glance this looks like a mass of difficulties where each must have a separate reason or cause, the fact is that all of these problems can stem from a set of experiences that the dog had, or didn't have, well before he was six months of age. It was the absence of these experiences in the final generations of the army superdogs that ruined them psychologically. Most of these problems can be avoided by the average dog owner with a little bit of work that starts when you first get your puppy.

The Chinese have a proverb that goes "A journey of a thousand miles must begin with a single step." On our journey toward creating the superdog, we began with the early stimulation program. It is a good first step, but only a single step, and, as we saw in the case of the biosensor dogs, if it is not followed by other steps, we will not make it to our goal, which is to produce a dog with a good, stable personality. The early-stimulation program is only the first of three components needed to produce the superdog. These are: early stimulation, socialization, and environmental enrichment. Please remember that for the purposes of this discussion we are not defining a superdog as one that flies through the air with a cape or can carry vehicles in his strong jaws. Rather, we are looking for a dog that you would welcome as a family member or neighbor, one that can socialize with other dogs, cope with the stresses of modern living, is accepted by the wider community, and is not overly emotional. We are looking for a dog that learns what you want to teach it and has a desire to work with you.

Critical and Sensitive Periods of Development

Before we go any further in describing the program to produce the perfect dog, you should understand that it is useful to look at the dog's developmental progress from birth through to adulthood as if it were not a continuous process, but rather as if it proceeds in a series of distinct stages. This was the conclusion of J. Paul Scott, who did research with Jackson Laboratory and was director of their dog research facility at Bar Harbor, Maine in the 1950s. Scott gathered around him numerous specialists on genetics and dog behavior, such as John L. Fuller and Michael W. Fox and others. From their research emerged the notion of a "critical period" in the development of any animal—defined as a special time in life when a small amount of experience will produce a large effect on later behavior. The difference between the amounts of effort needed to produce the same effect at different ages is a measure of just how "critical" the period is. For example, if a small amount of experience will easily produce a behavior at around three weeks of age, but at a later age, no amount of experience will establish the behavior, then we are looking at a rigid, inflexible, and definitely critical period. Different periods of development may vary in their critical effects. Suppose that we have a situation where at an early stage a small amount of experience will cause a behavior to appear, but at a later age, by investing a lot of patient effort for hours, weeks, or months, we can eventually reproduce the behavior that would have developed during the early period. Obviously, then, that earlier period of time is still very important but probably should not be called "critical" since some changes can occur after it (although with difficulty). In cases where learning at one age is very easy but there is still some small degree of flexibility later on, scientists have come to prefer the label "sensitive periods." When it comes to the development of young dogs there appear to be some important critical and sensitive periods that shape an adult's personality and behavior.

I will talk about developmental stages and critical periods as if they were distinct, with clear-cut boundaries between periods, but this is not really the case. Some of the stages overlap to a degree, and there is certainly a lot of overlap in the processes that are taking place in the growing dog. Thus, in human development it is useful to think about the stages of infancy, childhood, adolescence, and adulthood. However, if you were asked to define the exact ages that mark the beginning and end of infancy and the beginning of childhood, or the end of adolescence and the beginning of adulthood, you might be hard pressed to do so. This is because some individuals grow and mature at different rates than others. The stages in dogs are somewhat clearer than these are, but there are still individual differences.

The first of the dog's developmental stages is the *neonatal period*, which lasts from birth through to about two weeks of age. It is the period when the puppy is completely dependent upon its mother to keep it alive, since it is still blind and deaf, not very mobile, and helpless. For humans seeking to create a superdog, this is the period when the early-stimulation procedure should begin. You can start the process of stimulation when the pup is as young as two days of age; generally speaking, the earlier you begin, the greater the benefits in terms of neurological development during this phase. If you wait beyond two weeks of age to start, the improvements will be measurably less; however, since some gains can still be made this would qualify as a sensitive period.

The second stage is the *transitional period,* which lasts about a week, from the age of thirteen to twenty days. It really represents a transition from total helplessness to a point where the puppies can process information and voluntarily interact with the world. The most important aspects of this transition are that vision and hearing are beginning to function. In fact, we can mark this stage as beginning with the opening of the puppy's eyes at around thirteen days. You can tell that the eyes are functioning at this time because shining a light into them causes the pupils to contract, indicating that the light is being received and responded to. It may take a week or two more, however, for the puppy's eyes to respond to shapes and distances accurately and reliably. The end of the transition period is marked by the opening of the ear canals at about twenty days, which is when the pup will respond to loud sounds, like the clang of a metal spoon on a pot.

The first thing that the puppy begins to do once all of its senses are functioning is to try to learn about and interact with the world around it. The pup's growing awareness of his mother and his littermates quickly results in attempts at social exchanges and communication. At this stage, the pup will begin to move around more, wag his tail, bark, and even growl. You may even see the first evidence of play fighting with his littermates. As the puppy becomes socially conscious of his brothers and sisters, he will begin to view them as valued companions and not merely soft heat sources. Evidence for this comes from the fact that if he finds himself outside of the group in an unfamiliar environment he may yelp or whimper in distress, even if he is well fed and warm.

This is really the opening of the *socialization period,* which can be said to go from about three to twelve weeks of age. The term "socialization" is used to describe the process by which an individual learns about his social world. He learns what his family and the rest of the individuals who will interact with him socially expect of him. Thus, he is learning all of the rules and behaviors that allow him to become a functioning member of a particular social group. This process will actually begin during the transitional period when he makes his first attempts at deliberately relating with his littermates and his mother. The process will continue well beyond the twelfth week. However, we are justified at considering these nine weeks as a critical period, since what happens during this period may well have a greater influence than what happens at any other time of the dog's life. This interval fits the definition of a critical period better than any other does, since the presence or the absence of certain forms of stimulation, and the occurrence or nonoccurrence of certain events and interactions, during this window of time will shape the dog's behavior forever. If things go wrong, because of experiences the puppy has or experiences he misses during this period of life, the resulting behavioral problems may last a lifetime. The behavioral changes that result from what happens during these few weeks may be extremely difficult and perhaps impossible to modify later in life.

One sad example of the consequences of missed experiences at this age was told to me by a woman who owned a Jack Russell terrier that had been taken from her litter just a few days before it was five weeks old. This is far too early, and the dog did not learn the basics of social interaction with other dogs, including such basic communication as the "play bow," which is where a dog leans forward, with its front paws flat on the ground to the elbows and its hindquarters and tail straight up. This is the classic gesture that dogs use to invite other dogs to play. Never having learned the meaning of this signal, this unfortunate animal interpreted this behavior as a threat and responded by growling and charging dogs who were making friendly overtures. Her relationships with other dogs thus were brittle and hostile throughout the rest of her life.

Learning to Be a Dog

The aspect of socialization of a dog that springs to mind as our first concern is the process of learning to act appropriately around humans; however, this is only part of what must happen during this period to build a well-rounded personality. Although we call this the socialization period, three processes should be taking place during this time interval, namely: *identification* (where the pup learns that he is a dog), *socialization* (getting along with people and dogs), and *emotional control.* Some aspects of these will require human assistance if we are to end up with a superdog. Just as in the early-stimulation program, humans will be needed to control events, stressors, and other sources of stimulation that the dog receives.

It is important to understand that the problems facing a domestic dog are more complex than those that face a wild canine, such as a wolf. A wolf pup only has to learn that he is a wolf and then learn how to act around other wolves in a wolf society. Dogs, however, are domesticated and will live their lives with humans. This makes the process of socialization more difficult. Domestic dogs live in two worlds. They must first learn their species identity, which is of course that of a dog. This is obviously important, since the pup will have to get along with other dogs and function correctly in a canine society. While the dog must psychologically identify himself as a dog, he must also learn how to act and behave in a second society, that of humans. This is a delicate balance, since dogs must come to accept humans as part of their social world and must respond to human social gestures and behaviors, yet at the same time the dogs must maintain their sexual preferences for dogs and know how to act in ways that will still allow other dogs to accept them as members of the canine species. Simply put, dogs must be willing to accept both other dogs and humans as members of their family, pack, or society.

One of the greatest impediments to dogs recognizing themselves as dogs and knowing how to interact with other dogs is inadvertently created by humans. It happens when puppies are removed from their litter at too young an age, such as in the case of the Jack Russell terrier that I mentioned earlier. The problem is that it is only through social interactions with their littermates that dogs can learn that they are dogs. It is also through such interactions that they learn basic social and communications rules that will allow them to get along with other dogs. The window of time in which canines learn their identity and basic dog behaviors opens at about three weeks and closes between eleven and seventeen weeks. For a dog that is living in a very stressful environment (and that includes most of the wild canines), that window will close earlier, between seven and nine weeks.

Research from Scott's lab in Bar Harbor clearly shows what can go wrong. A litter of puppies was divided, with one-half of the litter being raised in complete isolation, with no contact with other dogs from birth through to the age of sixteen weeks. These pups were then reintroduced to their littermates who had been reared normally (in the company of the remaining littermates) and some other normally reared dogs (not from their litter). The results were not pleasant. The pups raised in isolation were attacked and rejected by the other young dogs. The problem was that the behaviors of the animals reared in isolation were all wrong. They did not try to socialize with the other dogs, and when approached by the normally reared puppies, their behaviors seemed hesitant, socially inappropriate, and inhibited. If, however, these pups were placed with a group of other dogs that had also been reared in isolation they did manage to live alongside each other without any attacks or other forms of aggression. This was not because they found solace in their shared unhappy experiences, since these pups still suffered the psychologically crippling effect that their early isolation had produced. They lived peacefully with the other pups that had also been isolated because they did not interact with them socially. For the most part they ignored their kennel mates, treating the other animals in the enclosure as if they were simply moving features in the environment, much like waving branches or leaves blown across the ground by wind, rather than actual living animals that formed part of a social group to which they also belonged.

Of course, total isolation from other living things is an extreme form of rearing that is seldom found. However, some breeders—out of a lack of knowledge or simply because they want to cut down maintenance costs—are willing to sell puppies at four or five weeks of age and allow them to be taken from the litter and given to a new owner. Such early removal from the company of other dogs has a damaging psychological effect that often seems similar to that observed in the isolated dogs reared at Bar Harbor.

If we look at the socialization period as roughly spanning the ages from three weeks to twelve weeks, then we can subdivide it into two separate stages. The period of three to five weeks of age is often referred to as the *primary socialization period.* It is primary because this is the period where the pup must learn that he is a dog, and he starts to learn basic dog behaviors.

During this primary socialization period, the pups can actually be socialized to any of a number of different species. I received an email that included photos of a mixed-breed terrier puppy named Flash, who had been rescued at just under four weeks of age and then reared in a home in Philadelphia. Because he was so young, his owners put him in with their cat Mildred, who had just had a litter of kittens, in the hope that the kittens would provide him with some company. Obviously, at this young age the pup had not yet formed his identity as a dog. Probably because Flash was the same size as the kittens and still carried the scent that most young animals have, Mildred adopted him as if he were one of her own. She routinely washed Flash with her tongue just as she did her own kittens. Flash could not know that he was not the same species as the other animals that were now his "littermates." Because his primary socialization was to cats it should not be surprising to find that many of his behaviors appeared to be more catlike than doglike. As he grew, it was clear that his favorite toys were cat toys, like a squeaky mouse and a ball with a bell in it. Because this age period is when the young pup learns social interactions and rituals, Flash learned cat mannerisms, such as the typical cat postures in stalking and pouncing when playing with his adopted feline family. He also learned certain other ritualized behaviors that are particularly feline, like the habit of washing his paws with his tongue and then using them to clean his face and ears. The proof that he was now socialized to cats and identified himself with cats is that when given a choice between spending time with puppies or with cats, he would always select the company of cats.

Flash's case is not all that unusual. A number of laboratory studies involving "cross-fostering" pups, which means placing them with a mother and perhaps young animals of another species, have shown that dogs can be socialized to rabbits, rats, cats, and monkeys.

Although learning how to get along with other dogs continues as long as the socialization window is open for domestic dogs, the period between approximately six to nine weeks seems to be when the pup most easily and quickly learns the social skills needed to communicate and meaningfully interact with other dogs. One of the most important lessons that the pup will learn during this time is *bite inhibition,* which means that the pup should have "a soft mouth." Specifically, the pup learns to mouth someone without biting hard enough to hurt the other individual or break his skin. This is an important aspect of communication.

Typically, a dog that is being annoyed or threatened will attempt to avoid direct physical conflict. This is an evolutionary holdover from when dogs were cooperative social hunters. Violent physical confrontations might result in one or both parties in the conflict becoming injured, and thus lower the efficiency of the pack as a hunting unit. To avoid such fights, there is a standard series of signals that escalate and tell another individual to back off and avoid direct confrontation. At the first level, the challenged animal may simply snarl or growl. If this is not sufficient to cause the threatening animal to back off, then the dog might move a step in the direction of the challenger and snap his jaws, making sure that he misses the other individual. If this "air snap" does not stop the threats, then the third stage of escalating aggression is an inhibited bite. In this case, the animal actually makes physical contact with his challenger, but controls the bite so that no puncturing or tearing of the other's skin occurs. Such an inhibited bite is usually enough to end the confrontation, since it clearly anticipates and emphasizes the final stage which is all-out, no-holds-barred, biting and tearing at the challenger with all of the attendant possibilities of major injuries to both parties.

The puppy first learns bite inhibition from his mother. This happens when he bites too hard on her nipple while nursing and is punished for his transgression. Lessons in inhibition of bite strength are repeated when littermates respond negatively when his bite hurts them. A littermate will obviously end any play activity if he is bitten too hard and may respond to hard bites by biting back. It is through such social blunders that the pup learns the importance of controlling the strength of his bite. There is good evidence that puppies that have been weaned too early or removed from their litter at too young an age will tend to bite with greater force and with less provocation than those that have been kept in the litter until around eight weeks of age. Such dogs often do not follow the usual sequence from growl, to air snap, to inhibited bite, and finally to uninhibited biting, but rather tend to move directly from growling to full-force biting. This is part of what appears to be a more general problem that such early-weaned dogs have in their ability to communicate. Not only do they fail to send the appropriate signals to other dogs, but they often do not correctly interpret the signals that other dogs send to them.

Dogs removed from the litter when they are still very young (four or five weeks of age) are much more likely to develop aggression problems toward other dogs when they are adults. The likelihood of such overt aggression toward other dogs steadily diminishes when the pup is kept in the litter until the age of seven or eight weeks. Alternatively, dogs that have been weaned too early may develop, instead of aggression, a pattern of extreme fearfulness when around other dogs.

If the dog is removed from the litter at too young an age, before it has identified with and become socialized to other dogs, then their human caretakers may become their major social focus. At first blush, you might think that it is a good thing to have a dog so focused on its owner, but this can lead to a number of attachment-related problems when the dog grows to be an adult. Such dogs may become extremely suspicious or aggressive toward strangers, perhaps because they view them as threats to their owners or potential rivals for their owners' affection. Such dogs will often act much the way that overly protected human children do, by becoming extremely dependent upon their caretaker—to the point where any separation from their human companion results in severe anxiety. This anxiety can lead to problems such as excessive barking or what appear to be compulsive destructive behaviors, whenever they are separated from their masters.

Learning About Humans

The period from six to twelve weeks of age is when the dog must fulfill the second requirement of being a domesticated animal: he must become socialized to humans. This time interval is usually referred to as the *secondary socialization period.* Notice that this actually overlaps with the second half of the primary socialization period of three to nine weeks of age.

Many insightful dog breeders and dog experts talk about the age of forty-nine days or seven weeks as the ideal time to begin a transition from living with dogs to living with a human family. In reality, there is a time period that extends from six to eight and a half weeks of age that works quite well. During the early part of this time, mothers begin to wean their pups. Since the puppies have not lost interest in their mother as a food source, this process can be a bit abrupt and harsh. The process generally reaches a climax when the mother snaps at a pup who has tried to put his sharp little teeth on her no-longer lactating teats. For the vast majority of litters, the period between seven and eight weeks of age is the time when the mother's job is finished—both nutritionally and psychologically. Although the pup may still be learning about living with dogs, this is a good opportunity for humans to step in and fill the social void left by the mother. In so doing the dog will begin to learn how to get along in human society, to respond to human communication, and come to accept humans as significant and high-status members of the community in which they will live for the rest of their lives.

Again, it was the Jackson Laboratories at Bar Harbor that conducted the classical research that established this principle, in a study referred to as the "Wild Dog Experiment." It got its name because the procedure involved rearing litters of puppies in large, open fields, where they had the regular company of dogs (their mothers and littermates), but only limited contact with humans. To see the effect that the dog's age would have on its socialization to people each puppy was given only one week of interactions with humans. During this socialization week, each dog would be brought into the laboratory every day where the humans involved in the project would handle it, play with it, and talk to it for a while. Some puppies got that week of human contact when they were two weeks old, while others got it at three, five, seven, or nine weeks. One final group never got any human contact or interaction until it was tested at fourteen weeks.

During the socialization exercises, the researchers noticed that puppies differed in their apparent interest in humans. The youngest pups, those that received human contact at two or three weeks, were only mildly attracted to the person who was trying to interact with them. Mostly they simply sat quietly near the researcher. The pups that seemed happiest and most comfortable interacting with humans were those that were between five and seven weeks of age. The pups whose socialization to humans began at nine weeks seemed noticeably more skittish or fearful when first approached by a person.

Testing was conducted at fourteen weeks. The puppies were put on a leash and taken on a walk through the laboratory building and up a set of stairs. For some, negotiating the stairs appeared to be a harrowing experience that left them frightened and trembling. The researchers observed and recorded each puppy's behaviors on this walk.

During such tests some pups show evidence of being alarmed and fearful because they are in a new, strange place. Being on a leash for the first time can make this more upsetting since the leash prevents the natural behavior of dogs, which is to run away and hide when they become frightened. When a puppy walks comfortably on leash and stays near the person holding it, he is actually demonstrating that he is comfortable and feels safe with the human nearby. Such cooperative behavior when on leash also shows that the pup is willing to accept people as leaders and to interact socially with people by watching them and following their movements. Puppies that are not comfortable in their social interactions with humans are not cooperative. They won't follow the person holding the leash and will balk or resist attempts to move them. Their resistance is usually greatest when passing through doorways, when out in a wide-open area, or when urged up a stairway. They demonstrate their lack of acceptance of humans as leaders by fighting the leash, running as far away as the tether will allow in their attempt to escape human control or by giving vocal evidence of their discomfort by whining, whimpering, howling, or growling in response to being restrained by their connection to the person.

It turns out that the age at which the pups had a chance to socialize with humans was the critical factor in determining how well they accepted humans during the test period. The puppies that had their contacts with people between five and nine weeks had the fewest problems on the leash. These pups followed fairly happily or required only a bit of coaxing. They resisted or balked only occasionally at doorways or stairs, but even then, they were easily calmed by the researcher and afterward continued on with no problems. The puppies that had their week of human contact before five weeks of age, when the secondary socialization period starts, were considerably less controllable on the leash. Waiting too long before starting socialization was even more harmful, and the pups that had had no human contact until testing at fourteen weeks were the worst. The puppies that had their human socialization too early (before the puppy's psychological condition was ready to accept socialization to a different species) or too late (after the critical period had passed) also showed the greatest levels of fear and anxiety by trying to escape contact or restraint, and whining and whimpering fearfully. When offered a treat by the researcher at the end of the leash test, they were often too overwhelmed by their anxiety to accept and eat it. Some even tried to snap at the researcher's hands when the person reached out to touch them. This research and other studies that followed showed that the most critical period for socialization to humans can start as early as around four weeks after birth and that socialization must have begun before the end of the twelfth week if it is to have any significant behavioral effects.

The disastrous effects of delaying beyond twelve weeks when this critical period for socialization to humans closes can be seen by what happened to dogs that had no human contact for fourteen weeks. They were frightened and uncontrollable. One of the researchers at the Jackson Laboratories decided to see if intense and continuous human contact could ultimately bring about socialization in such a dog using the same kinds of procedures researchers used to try to tame wolf pups. He took one of the dogs that had not had human contact before fourteen weeks, tried to make it a pet, and attempted to get it to socialize with humans.

When the dog was taken home and introduced to his human family, the pup was too fearful to respond to any human interactions. Like a wild-born wolf, it tried to hide and escape human contact. So the researcher resorted to using the same procedures used to tame wild animals. First, the puppy was confined so that it wouldn't run away. Next, the researcher began to hand-feed the pup, which forced this young dog into human contact for each mouthful of food. Eventually the researcher and his family were able to calm the dog and get it to accept close contacts with people, at least with people that it knew well. The unfortunate fact is that even with all of this intensive care, attention, and social contact, it never became a well-socialized dog. For the rest of its life it proved to be difficult to control and would not respond to human commands or human leadership willingly or well. It remained timid and fearful around strangers, and would snap aggressively if approached too quickly. Furthermore, as would be expected given the fact that it had been socialized to dogs but not to people, whenever it was given a choice to remain in the company of humans or in the company of dogs, it would always choose the dogs.

The Process of Socialization

Fortunately for those of us who want dogs with good personalities, the importance of puppy socialization is now generally recognized by the community of dog breeders. Good breeders handle and play with their pups when they are young. Better breeders actually bring in other people and children to touch and handle the pups and thus put their young dogs on the road to good socialization with humans. This raises the question, How much socialization is necessary to create a dog with a sound personality?

We can get part of the answer from adopting the perspective of the world as it appears to a young dog. In the puppy's view of the world, there are many different varieties of human beings. To a young puppy each type of human appears distinct and, during the socialization period the pup must learn that this broad range of creatures are all human and should be responded to in a positive way. To the pup a man and a woman look, sound, and smell differently. Children of various ages move, sound, and act in their own peculiar ways as well, from the awkwardness of the toddler to the noisy, quickly changing responses of the grade-school child. The pup needs to experience the whole range of children so that he treats all as people. Men with beards, people with hats, sunglasses, overcoats, umbrellas, necklaces, jangling bracelets, scarves, or ties must each be dealt with separately. People of different races, those with perfume or aftershave lotions, those who have recently eaten garlic and so forth, must all be encountered under safe, controlled conditions. People with canes, crutches, and wheelchairs may each also provide a new socialization opportunity.

It can be time-consuming to properly and fully socialize a puppy to people, but it is time pleasantly spent. Most people are willing to help since puppies are cute and appear quite safe. People naturally gravitate toward puppies and want to pet them and say hello. I have found that carrying a pocketful of treats that the pup's new "friends" and acquaintances can each give him will focus the dog on the people and ensure that they are considered to be worth his attention. Getting frequent treats will also keep his stress level down.

Some researchers, such as veterinarian and psychologist Ian Dunbar, suggest that it is important to get as much socialization as possible through contact with people. At one dog behavior workshop he was asked, "How many people are enough?" He replied, "The pup should meet one hundred new people by the time he is twelve weeks old, and two hundred would be better." The questioner seemed quite astonished by this statement and asked how this could be done. Dunbar smiled and said, "Use the puppy as an excuse to throw a lot of parties. You can also use it as an opportunity to renew your acquaintances with family or friends you haven't seen in a long time."

While the thought of having lots of puppy parties is appealing, one must be careful not to overwhelm the puppy with stimulation at this stage. Physical or psychological traumas suffered during this period can have lasting effects, so make sure the greetings and interactions with humans and other dogs are pleasant; avoid situations that might be frightening. The actual research in this area suggests that the amount of human contact that the pup needs is not very much. Some studies have shown that as little as five minutes of human contact each day throughout the four- to twelve-week critical period can achieve adequate socialization, but it is important that the puppy have human contact on a daily basis. To be effective, however, the time must be well spent, which would include social interactions with lots of talk and physical contact. Concentrated, active social interactions produce the best overall effects. Confirming Dunbar's general notion, the research suggests that increasing the amount of time with humans makes puppies more confident around people and forges a stronger emotional bond with the humans that will be its caretakers and family later in life. Increasing the number of people that the puppy encounters makes it less fearful of strangers in general.

The fact that dogs can become socialized to species other than dogs is a vital factor in our ability to control and work with them. It is also important for some other important functions, such as guarding herds of sheep. Typical breeds used for this purpose are the Maremmano, Great Pyrenees, and kuvasz. Behavioral scientists once thought that herd-guarding behavior, like herding, was genetically prewired. Research by Raymond Coppinger, a professor of biology at Hampshire College, has shown that that is not the case. He demonstrated that many of the herd-guarding breeds, even those coming from long lines of successful herd guards, turn out to be quite incompetent as guards, either running away from the flock or attacking or otherwise harassing the sheep. What determines whether the dog will be a good herd guard seems to have less to do with genetics and more to do with the dog's early history and socialization.

To create a good herd-guarding dog you need to rear it, from the age of about four weeks, with the sheep that it will ultimately be asked to protect. Except for his food, which is provided by the shepherd, the puppy has to fend for himself among the sheep. His social interactions are mostly with sheep. He grows up with the herd and lives with it for the rest of his life. Because of his socialization, he accepts sheep as natural companions and "pack members." When a predator, such as a wolf or coyote, approaches the herd, the guarding dog will rush toward it. Whether this is due to an urge to defend the flock that the dog has now socialized to, or whether it is simply to check out this strange new animal, is unclear. In either case, the dog interrupts the predator's usual stealthy hunting approach. When confronted by the dog the predator might respond aggressively by snapping at the herd-guarding dog (which will cause the dog to act more hostilely if it encounters this species of predator again) or it might take flight and run away. Either outcome protects the herd.

The socialization of herd-guarding dogs is quite complex. They must socialize to three different species. During their first sixteen weeks, livestock-guarding dogs live, not just with the sheep, but also with one or two of their littermates, a few adult dogs, which usually include their mother and other breeds of dogs (such as collies) that actually do the herding. In addition, they must interact with the human shepherd. This gives the dog the opportunity to accept dogs, sheep, and humans as appropriate targets for social contact and social interactions. The trick is that this has to be done during the critical period that extends from three to twelve weeks of age for the socialization to be successful.

Socializing a dog that will live with humans does not really require a lot of hard work, effort, time expenditure, and advance planning. The basic requirements of socialization are simply safe, nonthreatening social interactions. Talking with your puppy, touching him, playing with him, asking him to respond to basic commands, like sitting before receiving a treat or being fed, are all that you need to do. If you take a light leash, hook it to a belt loop, and keep the puppy with you as you go about your daily routine—that is often enough. Inform the puppy that you are going to move by saying his name and then give a signal, like "Let's go." This will cause the pup to understand that you are controlling the situation in a gentle way and focus him on the importance of paying attention to humans in his environment, and to you in particular. In other words, simply treating your puppy as a companion that has access to most aspects of your environment during your daily activities will go a long way toward producing a well-socialized dog.

A Teenager in the House

The first wave of research suggested that the socialization period comes to a close at around twelve weeks of age and defined the next three months (up to about six months of age) as the *juvenile period.* It was originally thought that this is the time when the dog's social behaviors begin to take on adult form. However, we have come to understand that the end of the socialization period is not as clearly defined as we used to think, and there may even be certain predictable substages during which particular aspects of behavior are most likely to be shaped.

For many dogs, the period between ten and sixteen weeks of age is important in establishing behaviors associated with issues of social status. During this time, the pups will try to control social interactions and test how much they can influence and control the individuals that populate their environment. Since most puppies have already been sent to live with their new human family, this is when the pups will see if they can control the humans in their lives.

What happens during this period is important in shaping certain aspects of the dog's personality. To get through this stage successfully a pup needs structure, meaning that things should be done in a routine manner, with meals, walks, bedtime, and so forth occurring on a regular schedule. Changing the dog's living conditions can be disruptive. Furthermore, the humans in the puppy's life should all act consistently. For example, all should use the same words for common commands, such as "sit," "down," or "come." Although the use of lots of rewards is encouraged, there also should be a gentle insistence that the puppy respond to commands. Thus, if the puppy is not coming when called, use a leash to tug him gently to you. Some authorities suggest that letting the puppy "get away with things" during this period may give him an overinflated feeling of how high his social status is in the family group, which may come back to haunt you later when you want him to respond to your directions and commands.

Sometime during the ages of four to eight months your puppy may seem to forget everything that you ever taught him. He might stop coming when called, might seem to forget his house training, and might even seem to avoid or challenge you. This period is comparable to the early human teenage years. The dog is experiencing growth spurts that alter the way its body feels as well as the beginning of puberty with its huge hormonal surges. Like a human teenager, he may start acting a bit unpredictably and may be more difficult to control as he responds to these body changes and some of those new emotions and sensations that seem so strange to him. This is the age range where owners are most likely to contact dog obedience instructors for help. In truth, this is fortunate, since this is a good age to begin formal obedience training. The very act of training helps to strengthen the bond between owner and dog, especially if the training involves many rewards, such as treats, play, or petting. It also helps to establish and reinforce the social dominance hierarchy since, obviously, the individual giving the commands and dispensing the treats is clearly in control and thus higher in status than his four-footed student who is performing the actions and accepting the largesse from his owner.

Exercise care, however, especially toward the end of this period, not to frighten or intimidate the dog through harsh behaviors, punishments, or exposure to anxiety-producing situations. Fears learned at this age, and emotional patterns established during this sensitive period, may be very difficult to change later in the dog's life. Patience, rewards, consistency, and routines work for the puppies at least as well as they do for human teenagers. One advantage that we have in dealing with the puppy over dealing with a teenager is that we can always place the pup on leash and keep him with us as we move around the world just as we did earlier in his life. This helps to reestablish appropriate social behaviors by refreshing his early socialization experiences.

Nancy, a colleague at the university, came to me a while ago to complain that her seven-month-old Australian cattle dog, Roo, had turned into a "sociopath." She told me, "He just completely ignores me, and even grabs my hand with his mouth if I try to make him do something that he doesn't want to. I don't understand what is going on—he was the star pupil in puppy obedience class and was perfectly happy to obey my commands. I think that Roo is smarter than Mercedes [her one-year-old Cavalier King Charles spaniel], who did a lot worse in puppy class, but she's much more reliable than Roo now, and she never forgot what she had learned there."

Nancy had encountered an interesting quirk in canine development. We have learned that how long the socialization period lasts, how long the teenage rebellion phase lasts, or even if that rebelliousness appears at all, seems to differ among different dog breeds. The dog breeds that have a longer secondary socialization period are less likely to show the periods of rebellion and dominance challenges. The best indicator of what is likely to happen at this age is the degree of *neoteny* that the dog shows. Remember that neoteny refers to how puppylike a dog appears (and acts). Dogs that look more wolflike, with longer, narrower faces and pricked ears (like the Australian cattle dog, Roo), seem to have more sharply defined and shorter socialization periods. These breeds are most likely to resist authority and dispute the status of their human caretakers as they enter their adolescence. The breeds that are more puppylike, with shorter faces, large round eyes, round head, and floppy ears (like the spaniel Mercedes), seem to have longer socialization periods that don't end abruptly. These more puppylike dogs often seem to skip the teenage rebellion phase completely and may show very few attempts to exert social dominance during these periods of their development. In Nancy's case, all that was needed was to go back to basics and give Roo a sort of "refresher course," where he was hand-fed and had to work for each bit of food that he got by responding to a simple command, like "sit," "down," or "come." This gently reminded him who was in charge, and after six weeks of this routine he had worked through his "teenage rebellion" and once again became the obedient, attentive dog that Nancy desired.

By four to twelve weeks of age, socialization to dogs and people may have been accomplished but it is not yet fully set and can be improved. Dogs that are well socialized at twelve weeks of age, but are then removed from the regular presence of humans and other dogs during the juvenile period, can lose their socialization over time. A dog that was properly socialized but then denied opportunities to interact socially for a prolonged time will start to act just like a dog that was not socialized in the first place (although the chance of improving the behavior at a later date will be better). The socialization process needs to start during the critical period, but it must be strengthened by repeated social encounters until the dog is six to eight months of age if it is to be permanent. If you have any doubts about this, simply remember the state that the Biosensor program's superdogs were in when they reached the Lackland training facility. Their first twelve weeks had involved extensive socialization and early handling. However, this had been allowed to lapse as these dogs found themselves in a sort of "passive storage" in the army kennels, without systematic human contact. Thus, all of the gains of their early rearing were ultimately lost. Creating a superdog requires a continuing level of social interaction, but it is easy work and well worth the effort.

CHAPTER TEN

How Environment Shapes the Superdog

Before we can understand the final steps needed to turn our puppy into a superdog, we will have to consider the pet rats of a Canadian psychologist and maybe also the effects of educational television programs such as *Sesame Street* for young children. Earlier I mentioned that there are three things we need to produce a dog with a good temperament that will fit well into human society and family life, namely: *early stimulation, socialization,* and *environmental enrichment.* The first two components have well-defined age requirements. If we do not perform certain actions and insure that our dog has certain experiences during particular windows of time, then the opportunity is missed and the dog may show deficits in his personality and psychological soundness for the rest of its life. However, the third component—environmental enrichment—has no time limits. It can have an effect virtually from the time the dog is about three weeks of age (when all of its senses are fully operational) until he a graying senior citizen. It can improve the intelligence and stress resistance of well-socialized dogs and also help remediate some of the deficits that may remain if the dog was incompletely socialized or has lost some of its socialization because it was relatively isolated later on in life. Environmental enrichment can also keep an adult dog alert and aware, and help to offset some of the mental and emotional problems associated with old age.

The Canadian Connection

It is usually difficult to locate the precise source of any new stream of research. In science, ideas seem to float through many minds at the same time, and often different people, separated by geography or language, may all come up with an insight at much the same time. For the psychological breakthrough I am about to describe, however, we have a pretty good idea who started it and how. It began in the mid-1940s when Donald O. Hebb, who was a professor of psychology at McGill University in Montreal, combined some casual observations with a theory that had been slowly shaping in his mind. Hebb was a behavioral neuropsychologist even though the label had not yet been invented, and was interested in the question of how we learn. Several aspects of this problem were of great interest to him, such as how we learn to identify things that we see, how we link ideas together to form new concepts, how we remember the solutions to problems, or how we learn to perform new skills, like playing a piano or riding a bicycle. Hebb really wanted to know what was going on at the neurological level, maybe even at the level of individual neurons, but unfortunately our knowledge of the physiology of the nervous system was still fairly meager sixty years ago when he began this research. Without available physiological information, he had to begin his reasoning at a more conceptual level.

Hebb already knew that stimulation was important for normal development of some parts of the brain. This is certainly true for basic sensory functioning. Take the visual system as an example. If we deprive an animal of normal pattern vision early in life by rearing it in the dark or covering its eyes so that only diffuse unpatterned light gets through, then that animal may well never develop normal vision. If you look at sections of the brain that normally process visual information, you find that neural cells look strange in animals that have been deprived of normal visual experience. They have many fewer connections to other neurons. In newborn animals, there are also few neural connections in these parts of the brain, but the number of connections normally increases as the animal matures. The absence of appropriate stimulation seems to stall the animal at this infantile stage of development, so the conclusion was that appropriate visual stimulation is needed before such connections would grow.

Even though it would be many years before evidence had been gathered confirming that the same processes were in effect in areas of the brain that were not associated with the senses, it nonetheless seemed reasonable to Hebb that those parts of the brain that are important for other mental functions would also require stimulation and activity to grow normally and achieve good levels of psychological functioning. He believed that stimulation, and the neural activities that result from that stimulation, affected the later activity in the brain. First, it encouraged the growth of new neurons in the brain and the growth of new connections between neurons. These connections (the synapses) between neurons could also change neurochemically with time and as a result of stimulation. Hebb's notion was that each time a neural connection between two nerve cells became active it changed something about that connection so that this process became easier the next time. Neurons that are frequently stimulated at the same time will then develop fast and strong responses, and it will grow easier for one to stimulate the other. This can ultimately lead to whole groups of neurons in the brain that are triggered all at once when some kind of stimulation enters the brain.

Hebb's idea is that neural activity has much the same effects on groups of brain cells as exercise has on muscles. A particular muscle that is exercised frequently will become stronger, larger, and more flexible. The reverse can also happen; namely a muscle that is not used will grow weak and might atrophy. Hebb believed that something similar could also happen in the nervous system, meaning that connections between neurons that are seldom used will become weaker and might actually disappear, the way that a muscle might atrophy and grow weak if it is never exercised. This leads us to a concept of a very dynamic brain; one that is continually "rewiring" itself and changing its structure based upon its experience.

Hebb suggested that part of the rewiring of the brain involved the creation of new circuits that he called *cell assemblies.* These new circuits might actually serve as a means of coding information. In practice, cell assemblies consist of many thousands of cells, but the connections could develop in such a way that activating any part of the circuit would cause the rest of it to activate. If we imagine that each cell assembly represents a memory, idea, concept, or object, it means that our experience is causing information to be stored that can easily be switched on again as a unit. Thus, these cell assemblies can be used in problem solving, learning, and reasoning tasks, and can serve as the basis for forming larger, increasingly complex cell assemblies that store elaborate memories and those associated with certain skills. You get the sense of this if you play a musical instrument like the piano. You do not seem to have to remember each note separately in a well-learned piece of music; once you start playing it all flows out as a single unified sequence without requiring any conscious thoughts about the individual notes. The important idea at the core of Hebb's theory is that the brain is not a fixed, unchanging structure, but rather a continually growing and changing structure based upon our experiences and our interactions with our environment.

Hebb and the Family Rats

In 1949 when Hebb first introduced these ideas in his classic book *The Organization of Behavior,* there was precious little evidence supporting the theory. It would take more than three decades for the actual data to confirm his theories. New studies would eventually provide data confirming the changing nature of the brain. Other research would link aspects of the structure and function of brains to the nature of the environment that individuals were reared in and the stimulation they received from it. Science would also develop new tools to analyze neural activity, such as functional magnetic resonance imaging (fMRI) and positron emission tomography (PET), that would allow us to see the activity of what looks much like the cell assemblies that Hebb predicted.

The important thing to be derived from a theory like Hebb's is that any experience that encourages mental stimulation and learning about the environment should result in the building of cell assemblies and new neural connections. The existence of new neural material and these cell assemblies should make future learning easier. Hebb was able to provide a bit of data that suggested that this was true, and this is where the pet rats enter the picture.

One day Hebb took home a few lab rats and gave them to his children to keep as pets. The children played with these animals and let them run around and explore much of Hebb's family home. Obviously, the life these rats were leading and the environments they were getting to explore were a lot more complex and stimulating than the standard barren laboratory cage, which might include only some wood shavings to rest on, a water bottle, and a food tray. According to Hebb's theory, these animals should have been systematically adding new neural material and rewiring their brains in a more efficient manner because of their experiences. This should make any future learning and problem solving easier and quicker. Hebb proved this by later testing the maze learning ability of these pet rats and comparing their performance to that of their littermates—lab rats that had been raised in the standard boring cages, where they had little to do or explore and where there were no problems or interesting situations on which they could exercise their minds. Theoretically, the lab rats should have smaller, less efficient brains with fewer connections between nerve cells. The pet rats, which had been more richly stimulated by living, playing, and socializing with Hebb's family, showed themselves to be better and faster at maze learning and seemed more confident and emotionally stable when placed in the testing situation—exactly as Hebb had predicted.

Shortly after this first set of tests on the pet rats, some of Hebb's McGill University research associates repeated these experiments using dogs. They compared the learning ability of pet reared dogs (who received all of the stimulation and varied experiences that a typical family dog normally has) with dogs reared in barren kennels in the lab. Actually, there were two types of lab kennels, one in which stimulation was kept at an absolute minimum by rearing the dog completely alone, and a second, somewhat larger type of kennel that the dog shared with a littermate. The dogs reared as pets with plenty of family interactions and many experiences in complex environments learned tasks fastest and most efficiently when later tested. The worst performance was for the dogs reared alone in a kennel. The dogs who shared a kennel with another dog did a bit better than those reared alone, presumably because the other dog provided some social and environmental stimulation, but they were still a lot worse than those dogs that had been reared as pets.

There were other differences between the pet-raised and kennel-raised dogs. These went beyond learning ability and suggested that personality, as well as intelligence, may have changed as a result of the levels of environmental stimulation that the dogs received. The pet dogs seemed to be less fearful and considerably less stressed in the testing situations. They explored the testing environments with a good deal of confidence and never showed much in the way of negative emotions. The dogs reared by themselves, and to a somewhat lesser degree the dogs reared with only a littermate, seemed to be more emotionally affected by their experiences. The dogs reared with limited stimulation became frightened in new testing settings. They often reacted in a hyperactive, almost frantic way. Any minor pain or discomfort, such as when the dogs were strapped into a halter (so that they could be tested for certain types of problem solving) could cause extreme emotional responses. Some dogs' behavior was clearly driven by fear, while others would respond aggressively when they were handled by the researchers.

The Breakthrough at Berkeley

The next leap in our understanding of what was happening to the brains and behaviors of these animals came in the 1960s. While to the outside world the only thing that seemed to be happening on the campus at the University of California at Berkeley were activities associated with the social, sexual, and political revolutions of the hippie era and the Vietnam War, a quieter revolution was taking place in the psychological and physiological laboratories of Mark Rosenzweig and his colleagues David Krech, Edward Bennett, and Marian Diamond. They were out to directly test Hebb's notion that experience can cause concrete and observable changes in brain structure and that those changes can have an influence on later behaviors. One reason this work was so revolutionary was that many scientists believed that the brain and nervous systems of complex animals essentially had their complete complement of neural cells a short time after birth. The idea was that beyond infancy, *neurogenesis,* which is the process of growing new nerve cells in the brain, was not possible. You might be able to grow new connections between brain cells (although that too had not really been proven yet), and you could, of course, lose brain cells through injury, disease, or simply due to age, but adding new cells was not believed to be possible.

The Berkeley researchers wanted to see if Hebb was correct and if stimulation from the environment that an animal was raised in made a difference in their behaviors, abilities, and brain function. They were very careful in their studies. To make sure that the effects that they were looking at came from the animals' experiences with their environments, they only used rats from the same genetic strain. This is equivalent to some studies on dogs that restrict testing to only one breed to avoid genetically based differences. In addition, they further restricted their studies to males to avoid any sex-related differences and made sure that they used littermates born at the same time from the same parents to keep their genetic makeup as similar as possible.

When the rats were twenty-five days old they were weaned from their mothers and sent to live in one of three specially designed environments that varied in the amount of stimulation they provided. One set of rats was raised in a relatively stimulation-free environment, alone, in a small cage in a relatively quiet room, without much to look at or listen to. Although they had as much access to food and water as the other rats, they were living in a rather boring version of solitary confinement. A second set of rats had a bit more stimulation and was housed in a slightly larger cage in groups of three. By rat standards, this would be a small but sociable group. In addition to the social interactions with their cage mates, they could look out and see normal human activities going on around them in the lab. The third group, however, was provided with an enriched and exciting environment that could be viewed as the rodent equivalent of an amusement park. Their new living quarters had ramps, ladders, running wheels, swings, slides, tubes to crawl through, and various toys and objects hanging from the ceiling. There were always new objects to manipulate and explore, and these objects were changed several times each week to keep things interesting. This big environment was shared with a dozen or so other rats, so that games and complex social interactions could take place in this continuously changing setting. These rats also received additional stimulation from their human caretakers who handled them on an almost daily basis. Thus, these rats had a world full of friends and toys and other sources of stimulation to occupy their time and exercise their minds.

The rats that were reared in the enriched (amusement park) environment appeared to be much more intelligent when tested. They solved problems and learned things much faster than those reared in the other two groups did. This increased learning ability showed itself most clearly not on simple learning or problem-solving tasks, but rather in problems that are complex by animal standards. The environmentally stimulated animals were significantly better in learning and problem-solving tasks that required complex choices and sustained attention. The animals that did the worst were those that lived in a less stimulating environment.

As in the initial results from the McGill research, there also seemed to be personality differences that developed as a result of the different rearing conditions and degrees of environmental stimulation. These showed up as emotional differences between the groups. Rats reared in an enriched environment appeared to be more stable, less easily frightened, and better able to cope with, and rebound from, stressful events. Although these results came from studying rats, the general pattern of findings has been shown to be valid for many other species, including dogs, cats, monkeys, and perhaps even humans.

Hebb and the Berkeley researchers were working on the theory that the brain responds to environmental stimulation with physical growth, which in turn affects all aspects of the mental functioning, including learning, memory, intelligence, and emotional responses. They had proved their theory with the behavioral evidence, but needed actual physiological data to demonstrate that the brain was really changing its structure. So the next step was to examine the brains of the rats.

As predicted, there were differences in brain structure, some of which were quite large. The cerebral cortex of the rats raised in the enriched environment weighed, on average, about 5 percent more than the cortex of rats from the other environments. The effect of this is even greater than it appears to be, since the rats in the enriched environment were also slimmer by around 7 percent because of all the additional exercise they had in their larger, complex world. Thus, animals from the enriched environment had significantly bigger brains relative to their body weight in comparison with their littermates reared with less stimulation. Research has shown that the size of the brain relative to body weight is a fair measure of the intelligence of an animal.

Because these results were so startling, these researchers went out of their way to rule out other possibilities, such as the suggestion that increased exercise alone had increased brain size. In later experiments, they exercised animals in a running wheel, allowed them to have other forms of activity, gave them lots of companions but little chance for exercise, and other variations and combinations of these manipulations. In the end they concluded that a combination of intellectual and social stimulation was required to produce these increases in brain size.

The Changing Brain

The next set of insights came from the Midwest, where William Greenough was carrying on similar research at the University of Illinois at Champaign-Urbana. His work demonstrated that the neural circuitry was actually changing in the animals that were exposed to the enriched environments. Very specifically, the number of connections between neurons in the brain was significantly greater, exactly as Hebb had predicted. In fact, in one study Greenough found that the number of neural connections increased in the range of 25 percent to 200 percent, depending upon which types of neural connections and which sites in the brain his research team considered. Furthermore, in the 1990s, in a series of studies that involved collaboration with James Black, Greenough was able to nail down exactly what it is about the enriched environment that is most important. This research showed that the most critical element that must be present in the enriched environment to produce these neurological changes is that the environment must provide opportunities for learning and problem solving.

Specifically, animals that are put in a large environment that provides them with exercise and exploration will not show the same gains in brain size and connectivity as another group that has to solve mazes that change from time to time or wend their way through obstacle courses that change erratically every few days.

Recent advances by two neuroscientists, Fred H. Gage, of the Salk Institute in San Diego, and Peter S. Eriksson of the Göteborg University Institute of Clinical Neuroscience, take us to the final step. While we have evidence that the cortex gets larger and heavier if we exercise mentally, are we also actually growing new brain cells? Using a set of sophisticated techniques involving special microscopes and antibodies that lock on to individual cells and can tell us if they have divided, these researchers found that enriched stimulation from the environment actually causes the growth of new neurons in the brain. Most exciting is that these new brain cells are forming in an area of the brain called the *hippocampus,* a structure involved in learning and memory. It also has the task of organizing information as it comes into the brain before it's sent out to the cortex for further processing and storage. To prevent the additional processing and storage from swamping the existing neural storage capacity, we grow additional neurons in the brain and add additional circuits and connections to neurons already present—the biological equivalent of adding new memory capacity and disk space to a computer so that it can run more powerful programs. It is as though the very act of interacting with the environment, processing new information, solving problems, and learning new relationships changes our brain so that we can learn more and organize information better.

The hippocampus, however, has other functions as well. It is part of the limbic system of the brain and is greatly involved in emotional responses. For example, a structure called the amygdala that hangs off the end of the hippocampus is involved in both fear and aggression. Remember that one of the personality changes that was observed in the animals reared in the enriched and stimulating environments was a marked reduction in fearfulness and stress. Thus, it becomes clear that the places where the brain is changing as a result of increased levels of environmental stimulation are consistent with the nature of the personality and behavioral changes that we are seeing.

A Cure for an Aging Brain

Now here is the truly exciting aspect of environmental enrichment studies. Early stimulation and socialization must take place at precise times during an animal's life if these experiences are to have any lasting effects. However, the beneficial effects of environmental enrichment can occur at any age. Greenough demonstrated this by doing another series of studies using elderly rats that had been reared in the relatively deprived conditions involving standard laboratory housing and low levels of stimulation. Late in the lives of these animals, he transferred them to the enriched, amusement-park environment with ramps, toys, and other rats. Initially, as you might expect, these old rats were quite frightened by their new surroundings, but after a while, they learned that there was nothing to fear. Once that happened they began to explore, to climb the ramps and ladders and use the slides, swings, wheels, and rat toys. They also began to socialize and interact with the other rats in their new and complex world and seemed quite happy in their new surrounds. When their brains were later examined, they showed increased cortex size and weight as well as noticeable increases in the number of neural connections. Fred Gage used his procedures to verify that adults and the elderly are also growing new brain cells in their hippocampus, much like younger animals reared in an enriched and stimulating environment.

Further research showed that these changes in brain structure in adult and elderly animals that are given environmental stimulation late in their lives also produces the same increases in intellectual abilities and increased emotional stability that we see when young animals are placed in such settings. Although, as usual, much of this work has been on rats, cats, and monkeys, we have some direct evidence from dogs as well.

At the University of Toronto a team of researchers, including psychologist Norton Milgram, has been looking at the mental changes that occur in dogs as they age. In humans, age often brings reduced memory ability as well as slower problem solving and reductions in general learning ability. At the extreme, we have senility and Alzheimer's disease, which can rob us of much of our memory and intellectual abilities.

We don't know exactly why dogs or humans decline in their abilities when they age, but it appears that damaged DNA, faulty enzymes, free radicals, and the buildup of toxins and metabolic wastes cause cell deaths and reduced function in sensitive neural systems in the brain. The physical results of aging in the brains and nervous systems of dogs and people are observable; old dogs have smaller, lighter brains than young dogs. The change is quite significant, and the older brain might be up to 25 percent lighter. It is important to note that this change is not necessarily due to brain cells dying off. The evidence suggests that we are losing those parts of the nerve cells that connect with other nerve cells (dendrites and axons). If we could consider the brain as a complexly wired computer, it would be the same as if various circuits in the central processor simply stopped functioning because connections were broken. Neurologists refer to this as a "pruning" of branches that are no longer used or needed, much as one might prune a bush in the garden. For the most part, it is the loss of these connections that reduces the size, weight, and efficiency of the brain.

As we consider these age-related changes, it is obvious that size, weight, and number of connections in the cortex are also factors directly affected by enriched environmental stimulation. Those things that age diminishes are the things that environmental stimulation increases. Greenough, working with rats, and Milgram, working with dogs, both speculated that perhaps this means that enriching the environment of individuals could help offset the negative intellectual effects of aging.

The University of Toronto studies have looked at a variety of factors influencing the aging mind, including diet and medication, but some of the most interesting work simply involved changing the day-to-day experiences of old dogs. Milgram's group used old beagles in their research and provided them with what they called "cognitive enrichment"—a program to exercise the brain. Five to six days a week, these dogs were challenged with learning tasks and puzzles, such as finding hidden food rewards. Stimulation of this sort went on for a year. Afterward, the animals that received the enriched mental experiences showed better performance in a set of complex learning tasks than their littermates that had not had these additional experiences. This was not merely due to staving off the effects of aging in a way that reduced the rate of mental decline. In a number of cases, the dogs actually showed themselves to be more mentally fit than they were when they first were entered into the study. Milgram summarized his results this way: "We say that we can teach an old dog new tricks because it's possible to slow down or partially reverse brain decline. Some dogs in our tests definitely became smarter."

Enriching Your Dog's Environment

Enrichment experiences can come in many different forms. Usually the best forms involve exposure to a wide variety of interesting places and things that can provide novel, exciting experiences, combined with frequent opportunities to learn new things, solve problems, and freely investigate, manipulate, and interact with objects and environmental features. The data is unambiguous in showing that this leads to individuals who not only tend to be more inquisitive and more able to learn quickly and perform complex tasks, but also who are less fearful and emotional.

Given the strength of the data supporting the usefulness of environmental enrichment, it is not surprising to find that psychologists have tried to apply these results to rearing and educating people. It was well known that children who did poorly in school and scored low on intelligence tests in general were often the products of underprivileged and non-enriched environments. Few attempts had been made by these children's caretakers to stimulate their thinking; they had traveled little and often grew up without exposure outside of their residence and immediate neighborhood. Such children are often described as growing up with limited social interactions, low levels of parental involvement, and few books and toys. Often the bulk of the child's experience consists of hours spent near a caretaker who pays little attention to him, and the major source of stimulation is a television tuned to daytime soap operas and talk shows.

When the Berkeley data on environmental enrichment was beginning to become widely known, a group called the Children's Television Workshop, with the help of psychologists and educational researchers, put together a TV program designed to attract children and stimulate their minds. In November of 1969, the first episodes of *Sesame Street* aired, and soon Bert and Ernie, Kermit the Frog, Miss Piggy, Cookie Monster, Big Bird, and others became household names. A fast-paced barrage of information was provided to hold the interest of young viewers. There were lots of opportunities for children to participate and to challenge themselves mentally, as puppets stared directly out of the TV screen and asked them questions like "What number comes after three?" or "Can you sing the alphabet song with me?" or "Can you point to the thing that is shaped like a triangle?" A research team headed by psychologist Edward L. Palmer showed that many children did join in the activities, singing, counting, and pointing as they watched the show.

This limited dose of environmental enrichment, originally designed for inner-city kids, was shown to have noticeable positive effects. When they were tested for their school-readiness, the children who viewed *Sesame Street* scored more highly, not only for specific abilities, such as alphabet and number knowledge, body part naming, and form recognition, but also for more general thinking skills, such as sorting, classifying, and understanding relationships, than did their non-viewing classmates. Furthermore, long-term studies that followed children for a number of years found that those who watched the program during their preschool years were still showing benefits even in secondary school. Thus, in high school, kids that had been regular viewers of *Sesame Street* when they were much younger were earning better grades in English, mathematics, and science, generally had higher grade point averages, and were more likely to read books than kids who had not had that experience. Those who watched the program most regularly seemed to have the greatest gains. John Wright, of the University of Kansas's Center for Research on the Influences of Television on Children, summarized the findings by recommending *Sesame Street* as "part of a balanced breakfast" for children.

The proven educational value of such enrichment has resulted in versions of *Sesame Street* being shown in more than 130 countries around the world. It also changed the face of television programming for children, so that many contemporary programs, such as *Blue's Clues, Dora the Explorer,* and *The Big Comfy Couch* basically ask children to help their characters solve problems and puzzles, while other characters, such as the purple dinosaur Barney, frequently turn to the camera and directly ask viewers to shout out answers to questions.

Fortunately, we don't need complex, expensive television programming to provide environmental enrichment for dogs. In some sense you can view environmental enrichment as a sort of "socialization to places and things" rather than to humans and canines. The requirement is that as a pup the dog will be exposed to many different environments, objects, and problems under safe, non-fear-provoking conditions.

In the beginning, when the dog is still young, the different environments can simply be different rooms with different floors. For the puppy, a room with carpeting is different from a room with a bare wood, tile, or linoleum floor. The way the pup must walk and balance itself on a floor that offers less traction provides a learning opportunity and a variety of different sensations. Moving the puppy to different rooms for fifteen minutes or so each day and allowing him to explore safely, either in the company of his littermates or with a human companion, is mentally stimulating. As the dog grows older you can branch out, exposing him to different places outside the home, such as a beach, woods, parks, parking lots, school yards, tennis courts, public buildings, quiet suburban avenues, or busy city streets. All help to exercise his mind.

When the puppies are still young, additional environmental enrichment can be provided by simply tossing some toys into their living areas. Toys that squeak or roll are attractive. Toys should have different textures, such as a heavy knotted rope, a tennis ball, a rubber ball, and so forth. Toys can also be the equivalent of environmental features for the puppy. Boxes to crawl into or on top of, tunnels or tubes to go through, a low chair to crawl under or jump onto, items hung from the top of a kennel or a hook on the ceiling so that they swing freely but so that their bottoms are only a few inches above the ground, all can provide amusement, entertainment, and a chance to interact and manipulate things. There should be only a few toys present at any one time, but they should be changed daily so that the pups don't become bored with them.

Talk with the pups frequently. Handle them and touch them every day. When you leave the place where they are, turn on a radio. Change the station each time you leave them so they will become accustomed to different human voices, noises, and styles of music.

Once the dog has been weaned and for the rest of its life you can easily continue to enrich its environment and stimulate its thinking. If you make food the reward for solving problems and finding things this will keep the dog's motivation high. For example, some dog toys can be filled with kibble. When the object is rolled around, or knocked about it dispenses bits of kibble. If you are willing to put up with a bit of controlled destruction, you can put kibble or treats inside a cardboard box, old towel or rag, or crumpled plastic jugs and allow the dog to tear the item apart to get to the food inside. The cardboard rolls that toilet paper and paper towels come on are great for this. Put some kibble in them, crumple the ends, and let the pup tear apart the "toy" to get to the food. Many dog toys are hollow, such as Kongs and hollow nylon bones, which can be stuffed with a dog biscuit, peanut butter, or cheese. The dog has to work at getting the food out. If you moisten some kibble, stuff the toy, and then freeze it, by the next day you will have a food-stuffed toy that the dog will have to work at for quite a while before getting treats out.

Empty water or soda pop bottles or milk jugs made of either cardboard or plastic make good toys. They are light and can be knocked around. They make noises when the dog closes his mouth around them and can be made even more attractive by stuffing a dog biscuit inside. Always remove the plastic rings and the plastic caps before allowing dogs to play with these items.

Turning meals into searches can also be useful. Dividing the dog's meal into small portions, each in a plastic container and hiding them around the house can keep a dog actively searching for a while. If you are willing to put up with the potential mess, simply toss some nonmessy bits of food around a room or yard and encourage the dog to find them. I like to freeze chicken bouillon in an ice cube tray and then toss out the chicken-flavored "pupsicles" so that they slide behind or under a piece of furniture while the dog watches. This then poses the problem of how to dig them out. Be careful, however, not to make the task impossible, or you will frustrate the dog and raise its stress level.

All forms of games based upon hide and seek are good. If you have someone to help you (I find visiting grandchildren are great for this), one person can hide and the other then encourages the dog to go and find them with a command like "Find Cora." In the beginning, the person who is hiding might have to call the dog from someplace out of sight. Once the dog finds the hidden person, he gets either a treat or a toy to play with. You can actually play a form of "canine tennis" with the dog serving as the ball: send the dog back to find the first person (who has now moved to a new place) and he sends him back to the second person (who has also moved their hiding place) and so forth.

Alternatively, you can hide a toy for the dog to find. Again start with a simple problem, by putting the toy on a chair or sofa in the same room, and then raise the difficulty level, such as placing it behind the chair. You can raise the level of complexity as the dog gets more proficient, by hiding the toy down the hall, up the stairs, in another room behind a piece of furniture, and so forth. Getting children to play this game with the dog (who is rewarded either with play or food for finding the toy) can keep both occupied for extended amounts of time. The children don't have to be told that they are enriching the dog's environment and building new brain cells and connections—they are simply having fun.

If your dog spends time outside, you must recognize that the average yard is usually a fairly boring, barren environment except for interesting things that might pass nearby on the other side of the fence. You can make this environment a bit more stimulating by hanging ropes or inner tubes from a branch or some other elevated item in the yard for the dog to play tug with. Change the terrain a bit by adding some big boxes that can serve as tunnels or platforms for the dog to climb on. Lay down small logs and lengths of PVC pipe (perhaps 5 inches, or 13 centimeters, in diameter) for the dog to walk and jump over while playing. If you have more than one dog, some barriers to hide behind or enclosures to hide in are useful, and the dogs will often create their own games using them. A child's wading pool with some water or sand can provide additional chances for play and interaction. Changing things in the yard frequently will also provide stimulation. However, generally speaking, the dog is apt to find the environment in the house, where people are moving about, more exciting than the average yard, so if you want the dog outside you should go out there to play with it occasionally.

Activities that are more formal can also provide mental stimulation. Dog obedience classes are not only a means of controlling your dog, but also are opportunities to meet new people and dogs in new situations. It is also a well-structured way to exercise their minds as they try to learn to respond to commands. Any dog sport activity, such as agility, fly ball, tracking, Frisbee, herding, earth dog trials, or field trials are also forms of enrichment as well as fun or useful things for your dog to learn. Teaching your dog tricks and games will keep him learning and continue the process of increasing the mass and complexity of his brain structure.

You do not have to devote your life to enriching your dog's environment and experiences. A few extra walks per week where you choose walking routes that are different each day will help. Taking your dog with you on shopping trips is one way to add novelty to the dog's life. Even if stores don't allow the dog to enter, you can leave the dog in the car for a short time, and when you have finished shopping take the dog out for a walk in this new environment, which is probably quite different from that near your home. Some people are lucky enough to be allowed to take their dogs to work occasionally, which is another opportunity for enrichment.

Teaching and training the dog are good ways to keep that canine mind active and solving problems. You can teach the dog new obedience commands if you like formal training. However, I simply like teaching my dogs words. I introduce the words in association with activities, and keep a constant patter of talk directed at the dog. Thus, the dog is told "In your house" when I want him to go into his kennel, "Office" when I want him to go into my office and wait for me, "Toy box" when I want him to take a toy from my hand and put it in his toy box, plus a variety of silly responses to words to keep me and my grandchildren amused, like "Turn around," "Back up," or "Find my hat."

The important thing to keep in mind is that your dog will benefit from increased environmental stimulation at any age, so this should be a continuing aspect of your life with your dog. Remember that while you are having fun with and interacting with your dog you are also improving his personality and strengthening his intellect so that he can be a superdog.

How to Create a Superdog

So let's put the whole Superdog Program together in a simple form. Remember the three keys are early stimulation, socialization, and enrichment. Now we can simply lay out what you need to do in a stepwise form.

- *Neonatal Stage* (birth to twelve days)
    - Biosensor program of stimulation
        - Hold head erect then point down (twice each).
        - Stroke and pet the pup while it is belly up.
        - Apply minor cold stress on towel or cold hand.
        - Apply touch stimulation between toes.
- *Transition Stage* (thirteen to twenty-one days)
    - Biosensor program continues.
    - Talk frequently with pups.
    - Begin introducing toys and novel objects into their environment (a knotted towel, empty plastic soda bottle, or cardboard tube works fine at this stage).
- *Primary Socialization Period* (three to five weeks)
    - The biosensor program of stimulation should continue through week 4 and perhaps week 5 if the pups are not too large.
    - It is important that the pups spend lots of time with their littermates to get their canine socialization and develop bite inhibition.
    - Start introducing the pups to friendly humans as part of socialization.
    - Expose the pups to different rooms with different floor surfaces for short periods (can be done with several pups at once).
    - Now and then briefly isolate each pup by placing him someplace alone for five or ten minutes.
    - Rotate toys and objects in the pen frequently.
    - Basic training (come, sit, and down) can be tried after the beginning of week 5 with lots of rewards.
- *Secondary Socialization Period* (six to twelve weeks)

- It is important that the pups continue to spend lots of time with their littermates up through week 7 or week 8 to continue their canine socialization and finish developing the emotional control associated with bite inhibition.
- Introduce the pup to as many friendly people as possible.
    - Try to get as many different types of people as possible (males and females, old, adult, young, different races, different dress style, et cetera).
    - A pocket full of treats helps, especially if the pup's new "friends" each give him one.
- Introduce the pup to many different places and settings.
- Change objects and toys in his environment frequently.
- Start positive training with simple commands (keep the sessions short and fun and use treats as rewards).
- Start encouraging game playing, like fetch.

- *Juvenile Period* (three to eight months)
    - Although you are finished with the most intense aspects of socialization, keep introducing your dog to people on a regular basis.
    - Now is a good time to starting taking the dog on short trips to novel places like sandy beaches, rocky shorelines, woods, et cetera, for enrichment.
    - Continue training your dog to respond to new commands and to understand new words.
    - It is important to maintain structure during this period. You want to maintain a predictable routine, to give the dog a feeling of control over his life, but to vary things enough to keep him from getting bored.
        - Teaching new games at this time can be useful.
    - If the dog starts dominance testing or becomes stubborn during this period, simply put him on leash, tie the leash to your belt, and keep the dog with you for a few hours at a stretch as you go about your daily business.
        - Having the dog work for each bit of food by responding to simple commands will help get through this "teenage rebellion stage."
    - This is a good time to start a formal obedience class.

- *The rest of the dog's life*
    - Enrichment should never stop.
        - Vary routines and toys.
        - Bring the dog with you on outings.
        - Include the dog in as many family activities as possible.
        - Get him involved in dog sports and games.
        - Keep teaching him new words, tricks, and commands.

The process of creating a superdog takes a little bit of work at first, but it is easily integrated into your everyday life. No matter how perfect your dog is, always treat him as a "work in progress" and continue to challenge him mentally. He will learn more, become smarter, become more stress resistant, and will be a better companion. It will also give you a sense of pride to know that you have shaped his personality as much, if not more, than his genetic endowment has.

CHAPTER ELEVEN

Creating Monsters

The woman's name was Janet, and she nervously stroked the head of a handsome gray and white American Staffordshire bull terrier named Ally. Janet's eyes were watery and she looked like she was on the brink of crying. Ally, however, seemed quite at ease as she stood up and moved close to me, lowering her wide head and rubbing my hand in an invitation for me to pet her. I did, but apparently she wanted more attention, since she sat up and put her paw on my knee. It was then that I noticed that Ally's toenails had been neatly coated in a very dark red nail polish with sparkles. Janet's mood and the tone of her voice were not sparkling, however, as she spoke about her problem.

"*You read about that pit bull attack on the little boy on a bike yesterday?*" she asked, and when I nodded, she looked down at Ally with an expression that was a mixture of sadness, concern, and perhaps even a bit a fear. "*The dog's name is Steel, and I was his breeder. He's from the same litter that Ally comes from. I just don't understand what happened.*"

She paused and wiped her eye. "*Look at Ally. She's a sweet dog. She's a certified therapy dog and we do hospital visitations. All of the dogs that I've bred have been like her. Ally's mother is the same way, and I made sure that the dog that I mated her with wasn't just handsome, but had a good temperament and no history of ever biting anyone. You know how afraid people are of pit bulls, and I wanted my dogs to be perfect—to be examples of why we don't need laws to ban my breed—my Ally.*

"Now even my own mother says that she is nervous around my dogs. She says if one dog in the litter is vicious, chances are that the others have the same bad streak. She says it's all in the genes, and I should be worried that Ally and the others will turn into killers without warning.

"She just doesn't understand. Ally is great around kids—loves them. She plays with my kids and their friends and has never nipped or growled or anything. That's why I don't know what happened with Steel. They tell me the little boy will be okay, but there will be scars on his face. Steel's owner had to beat him with a golf club to make him stop biting. I just don't understand what happened."

The problem Janet was facing is not unusual. There is a worldwide concern about aggressive dogs and aggressive dog breeds, which has occasionally resulted in so-called breed specific legislation and the outright banning of some types of dogs. Whether such actions are justifiable is a complex question, involving not just scientific data, but also social, cultural, and political issues. However, since the problems arise because of particular personality characteristics of individual dogs, it is important that we look at what science has to offer as an explanation.

If we have some control over the genetic makeup of a dog and have access to that dog from puppyhood, canine personality has enough flexibility that we can create a wonderful dog without an undo amount of effort and time expenditure. That very flexibility, however, not only allows us to increase the strength of traits that are desirable for a family or working dog, but also allows us to increase the strength of traits that make a dog unsafe and undesirable. That this sorry state can occur through ignorance of proper rearing and socialization practices is sad, but it is sadder still that some people deliberately manipulate dogs to make them vicious and dangerous. These are often the monsters that the media talks about after the deadly mauling of a vulnerable person or child.

Aggressive personalities are both born and made. It is possible to create an aggressive dog that is willing to attack other animals or people, even when it has not been threatened or challenged. People can create vicious, even ferocious dogs using variations of the selective breeding techniques and rearing practices that produce sound and agreeable dogs. Unfortunately, some individuals and groups have been purposely doing this for centuries. As a result, the personalities of some breeds, or certain bloodlines within a breed, have been distorted to such a degree that they are considered threats to public safety by politicians and the community at large. Consequently, some municipalities, states, provinces, and even some countries have instituted restrictions on or even barred the ownership of certain dog breeds.

The Pit Fighters

It is interesting to see how (and why) certain dog breeds have come to earn the reputation of being more aggressive than others are. Much of the impetus to create aggressive dog breeds is a direct result of the popularity of the so-called blood sports. In antiquity the best known of these were the Roman games, such as those held in the Coliseum, which usually involved gladiators in contests that pitted men against men in fights to the death, but often also pitted men against animals or animals against animals.

The first formally recorded use of dogs in the games began around 65 BC, with the Roman consul Lucullus, governor of Cilicia and later of Asia. The use of dogs in gladiatorial combat proved to be popular with the public so a quest began for larger, faster, more aggressive dogs. The Roman invasion and occupation of Britain was motivated, in part, by a desire to obtain the large powerful dogs that could be found there. These included the British mastiffs (which the Romans called *Britannias*), and also Irish wolfhounds and Scottish deerhounds. Many of these dogs were sent to Rome to die in the arena, forced into combat against lions, bears, bulls, elephants, and sometimes two-legged gladiators.

The Romans introduced their blood sports into Britain during their occupation of the island, although the spectacles there were on a smaller scale. The British developed a taste for such sport and long after Rome had fallen, bull and bear baiting continued as popular entertainments. In these contests, a bull or a bear was brought into the "pit," which was actually an arena, cordoned-off area, or a dug-out ring. The animal was then tethered to a ring in a wall or a stake in the ground, giving it just enough freedom to defend itself. A dog, or several dogs, was then released to fight the larger animal. If the bull was forced to the ground, or the bear was killed, the dogs were considered winners. If the dogs were killed, maimed, or hurt so badly that they refused to, or simply could no longer, continue the battle, then the bull or bear was considered the winner. Meanwhile, bets were taken, not only on the final outcome, but also on whether a particular dog could seize the bull by the nose, or the bear by the neck, how long it would hold on, and how long the fight would last. As in current horse racing today, there were touts peddling the equivalent of tip sheets, with the previous records of the bull or bear and the dog or dogs listed to assist those placing bets.

These sports were popular, not only among the masses, but also among aristocrats. Henry VIII was a fan and had a pit constructed at Whitehall. Queen Elizabeth I was a great fan of bear baiting and even appointed an official Master of the Bears and Dogs who received a princely annual wage amounting to more than twenty times what a middle-class merchant might expect to earn. Elizabeth bred her own line of mastiffs for the pit sports and frequently would take foreign ambassadors to see contests. In 1575, for example, "a baiting display for Her Highness's pleasure" involved thirteen bears.

With the growth of public sentiment for more humane treatment of animals, and to curb some of the criminal activity associated with the theft of animals to be used in pit fights, the British Parliament attempted to ban bull and bear baiting. Actually, because of the popularity of these sports they did not attempt to stop the pit sports completely, rather as a gesture of humaneness, they elected to ban holding such contests on Sundays so that people would be more likely to go to church and spend time in contemplation. Elizabeth was so incensed by Parliament's attempt to curb one of her favorite pastimes that she exerted her royal prerogative and overruled the edict.

Elizabeth's successor James I continued the royal tradition of supporting these blood games. His passion for such "sport" verged on being an addiction, and he was always seeking ways to make the fights more exciting. At one point, he came across a statement by the sixteenth-century travel writer Abraham Ortelius, in which it was claimed that the English mastiff was just as courageous as any lion. Finding that some lions were kept in the Tower of London, James ordered a combat between three mastiffs and a lion. The dogs were let loose against the lion one by one. The first two did not survive the combat; however, the third did manage to bite the great cat so severely that it sought refuge in a den that had been provided for it. The dog was then declared the winner and given a comfortable retirement (because the king felt that any dog that had successfully fought a lion should never be matched against any inferior creature). This dog was extremely lucky, since the most usual procedure, when a dog won a fight in the baiting pits, was to find another, stronger, opponent to match him against the next time. It was not expected that dogs would survive more than a few fights.

After the Ban

Over time, however, the brutality of the sport finally became an issue with the public. Some clergymen began to speak out, rejecting the earlier arguments of the church that there was no cruelty involved in these activities since animals had no souls and therefore could not feel pain. These clergymen also protested that children were being taken to such events and were learning that cruelty and killing were acceptable behaviors. Humane associations were formed to lobby Parliament, which resulted in the passage of an act outlawing baiting sports in 1835. While this eventually brought bull and bear baiting to an end, it actually served to promote the alternative sport of dog fighting.

After the ban, the owners of "bulldogs"—the name given to dogs used to bait bulls, bears, and other animals—began to stage fights, just between dogs, to satisfy the demand for blood sports and gambling. These contests did not require large arenas, but could be conducted in much smaller spaces. Temporary rings or pits were easily constructed and then torn down after the contests, so it was difficult for the authorities to intervene.

To better fight other dogs, the big, heavy bulldogs were crossed with smaller, quicker terriers to produce the "bull terriers," which became the basis for the most popular fighting breeds. Staffordshire bull terriers, American Staffordshire bull terriers, and pit bull terriers are all derived from this ancestry. It is because they fought in an enclosed arena called a "pit" that they are called pit bulls.

Dog fighting soon moved to the Americas spurred, in part, by the importation of the Staffordshire bull terriers, which began in 1817. Although for more than fifty years dogfights were very popular in the United States, resistance mounted against this cruel sport. At the forefront of this resistance was the author Jack London, whose novel *White Fang* tells the story of a wolf-dog hybrid who is stolen and forced to fight while spectators gamble on the outcome. London used the success of his novel to lend credibility to the Band of Mercy Children's Crusade's campaign against dog fighting, a group that had been organized by George Angell, who founded the Massachusetts Society for the Prevention of Cruelty to Animals in 1868. London's first goal was to eliminate coverage of dog fighting from the sports pages of respectable newspapers. He was successful and after results of dogfights disappeared from the popular press, it considerably diminished the image of dog fighting as a legitimate sport. Nonetheless, dogfights before crowds of rowdy gamblers continued to be legal in many American states for years after London's death in 1916. In the end London's campaign was successful, and early in the twentieth century most of the American states and Canadian provinces had banned dog fighting activities.

Macho Dogs for Macho Men

If dog fights are illegal, and merely a cruel sport that was deemed illegal nearly a century ago, why should we worry about them today? Unfortunately, dog fighting has not disappeared. It is big business and brings in big money. There are underground magazines (such as *Sporting Dog Journal*) that list fighting dogs and their records. Such publications and associated websites provide contacts where information about upcoming fights can be obtained. The internet has made it easier to contact others interested in dog fighting and to arrange matches. Websites and bulletin boards are hosted by servers in locations where dog fights are still allowed, hence avoiding legal problems. There are no certain statistics for how many dogs and people are involved in this business since it is an underground activity, but the Humane Society of the United States has estimated that up to forty thousand people are earning most of their incomes from dog fighting–related activities in America. One Chicago police officer, Sergeant Steve Brownstein, works full time on the problem. Over a three-year period, he has seized more than seven hundred dogs and made more than two hundred arrests—and this is only one officer in one city, which should give you an idea of how extensive the practice must be.

The popularity of dog fighting is actually increasing today. There are aspects of popular culture that seem calculated to suggest that this illegal sport is fashionable. The mass media contributes to this by popularizing certain stereotypes. For instance, pit bulls are viewed as tough dogs and are seen as the perfect companion for individuals who want to appear mean and macho. It has now become almost a cliché on television and in movies to see fearsome-looking pit bulls straining against heavy chains or snapping and growling from the far side of a chain-link fence as they guard the homes of their gun-wielding, drug-dealing, lowlife masters. Such dogs also appear in hip-hop videos, often with their ears cut to look like devil horns. In these, they can be seen snarling from the backseats of high-powered SUVs, providing a visual accompaniment to antiestablishment gangster rap music. In this context, the dogs are used as status symbols to demonstrate how mean and tough the rap singer is. Thus, one of the original gangster rappers, Ice-T, chanted, "I got a pit bull named Felony." It has gotten to the point that the meaner and more aggressive-seeming the dog, the more effective it is as a prop for the hip-hop singer and his "posse."

Unfortunately, looking tough is not enough. To confirm their status many people feel the need to prove how tough their dog is, which they believe can only be done by having it defeat other dogs in combat. Rappers celebrate not only fighting-dog breeds, but also the actual dogfights. Thus, the rap star DMX can be seen actively promoting and glorifying dog fighting in his video for the song "What's My Name," which takes place in a futuristic dogfight pit. In another song, "Dog Match," DMX teams up with the singer Eve, and they precede to rap about the bloodlines of fighting dogs, how dogs fight, and pups that can "stand a match for three hours at least." His album *Grand Champ* features a pit bull on the cover and is an homage to dog fighting as a sport. Similarly, a video for the song "99 Problems," performed by the rapper Jay-Z shows several clips of dogs fighting at an arena. Rap stars popularizing dog fighting have probably led to the increase of "casual" dogfights, staged in streets and alleys and often attended and even organized by young adolescents. In such street fights the stakes are more likely to be reputation and bragging rights, rather than large sums of money.

An Undercover Operation

Not all people believe in the humane treatment of animals. I feel that I should start this section with a cautionary warning, such as that used before some television broadcasts of violent and gory programs. The information that follows is graphic and demonstrates the tragedy of what happens when unscrupulous people breed dogs for an aggressive personality. I was surprised and appalled when I became aware of the extent of dog fighting in today's "civilized society" and feel that there needs to be a wider awareness of this problem.

My real introduction to the dark world of dog fighting occurred several years ago. I was contacted by a person (call him Brad, since he doesn't want his name known) whom I had met when he attended a dog behavior workshop that I presented. He had learned that I was going to be in Michigan for a scientific meeting. Brad explained to me that he was associated with the Michigan Humane Society as part of a campaign to gather information on a dog fighting ring that operated in and around Detroit, Michigan. He said he would like my help and asked if I would attend a dogfight with him. I was very hesitant because it was my understanding that simply attending a dogfight was a misdemeanor in Michigan, but Brad assured me that he and the society were in contact with the authorities who agreed that this information gathering was a worthwhile enterprise, and we would not be prosecuted. Furthermore, he said he needed my assistance because his "cover story" was that he had a Staffordshire bull terrier that he wanted to have trained to fight. The dog, however, was actually borrowed, and because it was an extremely strong and active animal, Brad was not sure that he could control it and appear comfortable with it in a noisy, chaotic setting. He and I were supposed to make contact with some individuals associated with the sport at which point he would talk to the people involved while I kept the dog under control. Our task was to try to identify the organizers of the fight, the suppliers of the dogs, and the people training the dogs for competition. I knew about dog fighting only through written descriptions, but it seemed like a cruel practice that should be stopped, so I agreed to assist him.

We arrived at the location of the match, a vacant auto repair garage. It appeared that more than one hundred people were gathered there, including around a half dozen young adolescents. We spoke to two men who seemed to be in charge. One was dressed like a biker gang member and the other was dressed casually, with good quality clothing, and looked like an average member of the upper middle class. This second man told us of a breeder-trainer, whom we'll call Dave, who was the person that he felt we should see. He pointed to a thin muscular man dressed in jeans and a denim shirt in the front of the crowd, close to the fighting ring. Since the fight was about to begin we would have to wait until it was done to talk to him, so Brad and I turned toward the center of the room where the action was taking place.

The pit was an area about fifteen feet square bounded by low plywood walls. It had a green-carpeted floor with a few dark brownish stains that I suspected were bloodstains that had not been washed out since its last use. The procedures of this modern dogfight differed little from those of the 1800s. The pre-fight ritual began with weighing the dogs. When arranging each match a weight is set that neither dog may exceed. Next, the dogs were washed with soapy water—both from the same bucket. This is the traditional safeguard to remove any drugs or poison that may have been placed on a dog's fur to harm its opponent. The rules of the fight are ancient: the dogs fight until one is too injured to continue or alternatively turns and runs in defeat, or rolls over submissively. At that point, the dogs are given a one-minute rest and then set on each other again. If the dog that "turned" still refuses to fight, he is declared the loser. The fight I saw lasted more than forty-five minutes, with spectators and handlers urging the dogs on. The sound in the room was sometimes deafening as the crowd jeered and shouted when one dog or another seemed to gain some advantage. All the while bets were being made, some of which seemed to be measured in thousands of dollars.

The dogs that were in this particular fight appeared to be Staffordshire bull terriers (like the dog that Brad and I claimed we wanted to train to fight) and they sprang out at each other with an intense ferocity but with surprisingly little noise. I was later told that how much noise a fighting dog makes depends upon how much of the heritage is from terriers. Dog fighters argue that the "terrier blood" brings out the noise during a fight. The dogs' strategies varied from moment to moment, and although sometimes they simply slashed at each other, it became clear that they were looking for a place to bite and hold. Once actually bitten, a dog would break free and immediately turn on its opponent. Within fifteen minutes or so, both dogs had wounds that were visibly dripping blood. Toward the end of the fight, one dog was clearly in bad shape. One ear had been almost completely torn off, and he appeared to have a broken leg. Finally, in a desperate attempt to fend off the attacks from the other dog, it tried to hide behind its master. As specified in the rules, this caused the fight to be paused. However, after the mandatory one-minute break, the dog was released again and despite its injuries, it staggered forward, only to be bitten in the chest and neck area by its opponent. It managed to tear free and again sought to use its master as a shield from the onslaught. This time, with blood running down its front and one leg useless, the dog would not return to the fight and was declared the loser. Its owner looked at it in disgust, cursed, and kicked the dog away from him. He said something to the "referee," then snapped a leash on the badly injured dog and dragged it from the ring.

Another fight was to follow, but during the break, while bets were being settled, we walked over to Dave and arranged a meeting for the next day. We walked out, and since everyone was still inside, we took the opportunity to copy down the license plate numbers of the vehicle we knew belonged to the biker who was one of the fight organizers, as well as a few others from vehicles belonging to handlers who had brought dogs to fight this evening. These would be given to the police in the hopes that they could identify some of the participants. As we passed by one pickup truck, I noticed that it contained the bloodied dog that lost the fight. He was lying on his side on a bloodstained tarpaulin. His ear appeared to have stopped bleeding, however, when I looked closer I saw that he was no longer breathing.

Creating the Monsters

The next day we visited Dave. He explained that he bred a fine line of fighting dogs. All came only from long lines of winners, and all of his puppies were guaranteed to be good pit fighters.

"*It's all about game,*" he explained. "*A game dog will go at anything and just keep going at it. It won't stop fighting no matter how much it's hurt. It's got heart, courage, and grit. You've got to pry its jaws open at the end of fight to get him to stop biting his opponent. That's what the breeders of Staffs* [Staffordshire bull terriers], *Am. Staffs* [American Staffordshire bull terriers], *and pits* [American pit bull terriers] *are breeding for. But you can't prove that the dog is game unless you fight him.*"

I later learned that Dave had given me the code phrase that would allow me to find people who were breeding dogs for fighting purposes. Enter "game bred" and "dogs" into your internet search engine and see for yourself. These terms will take you to forums and chat sites about dog fighting. It will also bring you to a variety of people who claim that they are selling "game-bred dogs." Certainly the names that some of these breeders use for their kennels, such as Combat Kennels, Fighting Machine Kennels, Hard Bite Kennels, No Mercy Kennels, or Seek and Destroy Kennels, do not suggest that they are trying to attract dog buyers looking for fun-loving, friendly house dogs. Although most of their websites have a disclaimer that suggests that they are not trying to supply dogs for "illegal activities," which should include dog fighting, they also go out of their way to advertise that their dogs "will not back down," "will stand up in any fight," or "can put down anything or anyone in their way." Such statements suggest that they know what their clientele wants to use the dogs for, regardless of their disclaimers.

Next Dave took us around to see the facilities that he used to train dogs for fighting. Lined along one wall were a series of wire enclosures containing rabbits, some cats, and a few small to medium-sized dogs of nondescript breeding. When I asked about these, he explained:

"You got to give the dogs a taste for blood. Rabbits are an easy way to do that, especially if you keep the dog a little hungry. Next you work up to cats and then other dogs that won't put up much fight—you don't want your fighter hurt before he's fully trained."

He then showed us something he called a "cat mill." This was a contraption with two rigid bars that revolved around a center point. One bar had a cage at one end and the other had a harness on it. A dog was placed in the harness and a cat in the cage, with the spacing between the bars set at a distance where the dog would try to get at the cat. As the dog raced around in a circle, it became more angry and frustrated, but also was getting physical conditioning from the running. At the end of the session, the dog got to kill and eat the cat as a reward.

"The hardest thing is to get enough dogs and cats to use as bait during training. One good source is to check the local newspapers for ads that offer a pet dog or cat 'Free to a good home.' The cats that you get this way are better than strays since many of them have had their claws removed, which makes them safe for pups to start with. My wife dresses up neatly and collects them for me. If that source is running dry, or if I start running out of bait animals because I'm training a big bunch of dogs all at the same time, I just hire some local gang kids to fetch me dogs and cats. They scour the backyards and alleys and take whatever number of animals we need. They make sure that they leave the doors or gates open, so people just believe that they accidentally left them unlatched and that's how their animals got out and got lost."

Treadmills for dogs and heavily weighted collars or heavy chains were also being used to toughen and condition the dogs. Automobile tires suspended from ropes were used to strengthen bite power as well. *"If the dog isn't interested in biting the tire you can always hang a cat in a sack with just its paws free. That gets them interested. Then later you stuff the sack and the cat in the tire. You should see them go at that!"* He laughed as he described these additional "training tricks."

As we left, Brad told me that the Humane Society would forward this information, especially the fact that pets were being stolen as bait animals and the names of the gangs or neighborhoods involved, to the police. I later learned that police action did not begin in earnest until nearly six years after our covert data-gathering expedition. Even then, the police action did not come about because animal abuse and dog fighting were of great concern to the authorities, but rather because the nature of the people who were organizing and attending the dog-fights became of interest.

Dogfights and the Mob

Professional dog fighting (as opposed to the more informal street fighting) is often associated with gangs and criminal organizations. We can trace the modern history of organized crime in America to the Prohibition era when illegal liquor manufacture and distribution became big business. After a while, the major organized crime syndicates diversified, moving away from the liquor traffic and into gambling, loan-sharking, prostitution, and drugs. Dog fighting was not yet organized professionally but was run by small local groups. Organized crime became seriously involved in dog fighting only in the last few decades. This appears to have started gradually, perhaps as an outgrowth of acquiring pit bulls and similar breeds as guard dogs, combined with the macho idea of proving that their dog is stronger than the others. The gambling aspect of dog fighting was also attractive to the mob. It was easy for mob-related groups to contact fans and gamblers, since many of the people who are interested in dog-fighting are also interested in guns and drugs. One investigation conducted into a dog fighting ring in Arkansas by the Humane Society of the United States uncovered the largest supply of illegal drugs, gambling money, and handguns ever seized in that state up to that time. It was the link between these illegal commodities and dog fighting that seems to have ultimately caught the interest of police.

In Detroit, for example, Commander Gerard Simon, of the Organized Crime and Gangs Division of the police, was engaged in raids looking for drugs and guns. He said, "We kept coming across the dogs and dog-fighting dens and we decided we had to do something about it." Police were accompanied by members of the Michigan Humane

Society (including Brad) and the city's Animal Control Unit. In one sweep they acted on thirty felony and misdemeanor warrants for people wanted in connection with dog fighting and animal cruelty. I was pleased to learn that the man we knew as the biker, and Dave, the trainer, were included in that set of warrants.

Commander Simon sadly observed, "Drugs, gangs, dope, dogs, and Detroit. They all go together." Unfortunately, he was being too hard on Detroit. Illegal dog fighting occurs in every state in the United States and in all provinces in Canada. It is a worldwide problem, and Germany, France, Italy, and Britain are all experiencing a resurgence of this blood sport. Furthermore, in many parts of Asia, Latin America, Africa, and the Middle East, the sport is not only practiced, but is legal.

The Personality of a Fighter

I raise the issue of dog fighting because successful fighting dogs require not only a particular set of physical attributes, but also a personality suited for the pit. You must psychologically alter a dog to turn him into a good pit fighter. While some of this can be done through the kind of abusive training that Dave showed us, it is easier and more efficient to do this genetically.

How do you produce the perfect fighting dog? You want a dog that is not only strong and quick, but also insensitive to pain, a dog that is willing to attack with minimal provocation, and that will continue to attack even if it is hurt or the other dog has given up the fight. These are the dogs that win matches in the pit. So obviously the program for creating genetically aggressive fighting dogs begins with the unconditional rule that only match winners will be bred, and then only to other match winners. Preferably, dogs that have survived many fights will be used since they will have the best combat attributes. Fighting-dog breeders like to boast that they keep the bloodlines "pure," which means that not only the parents of their pups have survived the purgatory of pit fighting, but also their ancestors had to have been successful fighters as well, for as many generations back as possible. This unconditional selection process has been in use for nearly two centuries, which is the secret of producing good fighting dogs. A biologist would describe this process as "performance-based applied behavior genetics."

The personalities of these fighting dogs are different from those of other dogs. This process of selection systematically breeds out the basic social responsiveness of the dogs that normally would make them cooperative, pack-living animals. Such dogs do not react normally to the signals associated with the social order of the pack. The game-bred fighting dog does not accept or respond to signals of submission from another dog that does not wish to fight or has given up the fight. It has no inhibitions about which dogs it will bite. This means that a male fighting dog will bite a female, a defeated injured dog, a young dog, even a tiny puppy. These fighting animals are easily stimulated to the point where they will launch an attack. Sometimes the stimulation that triggers their aggressive behavior is inadvertent or accidental. Thus, I know of a woman who was attacked by her pet Staffordshire bull terrier, simply in response to her slipping and falling close to him. She was merely lying on the ground trying to get up when the attack occurred.

This last case illustrates the dangers associated with the loss of the personality characteristics that define dogs as pack animals. Dave mentioned some of these dangers when we visited his training facility:

"You got to be real careful about these dogs when they get the fire in them. When they're angry and you try to hold them, or just get in their way, you're gonna get hurt. Me and other trainers have all been bit at one time or another. When these dogs are in a rage, they just seem to be looking for something to fight. If you give them another dog or some other animal, they'll go for it, but if none are around, they'll go at you. It don't matter if you're the human that has been feeding them, and training them, and should be the boss. Scientists have got that pack order thing all wrong. At least when it comes to fighting dogs, none of them accept anyone as their leader or what those scientists call the 'alpha.' That means that you got to be careful all of the time and watch their moods. Sometimes working with these dogs feels like trying to juggle loaded guns—if you let your eyes wander for a second, you stand a real chance of getting hurt."

This makes the process of breeding the descendants of fighting dogs difficult, mainly because the maternal instincts of many of these dogs have become distorted. Dave explained it this way.

"*Breeding these dogs can be a bitch—because of the bitch,*" he chuckled at his own verbal cleverness and continued. "*You got to watch the litter of pups all of the time, at least for the first few weeks after they've been born. Remember that that mother has the heritage of fighting dogs in her. If one of those pups acts up, like say, biting her nipple too hard, she might just kill it or hurt it real bad. Motherly love is not part of what a fighting dog has. Because of that, as soon as the pups are old enough to go on to solid food, we take them away from their mother for their own safety.*"

The fighting instinct and aggressive personality become evident early in these game-bred dogs. This means that puppies are not truly safe even after their mother is removed, simply because their littermates are also potentially dangerous. All puppies fight, wrestle, and nip at each other; but fighting dogs are different in their behaviors and the level of aggression that they show to their littermates. Dave described his experience with puppy behaviors this way.

"*Even when they are just weaned from their mother, these pups are already fighters. They are already developing that lock-jaw bite that they'll use later in the ring. You know, the puppies will lock their bite on one of their littermates so hard that you can lift up maybe five or so of them at the same time, with each of them locking their bite hard on the skin of the next one. The only way to break up that cluster is to put your hand over the nose of each pup, one at a time, to cut off its air and make it let go.*

"*They grow up with their fighting skills. At about two and a half to three months of age, the pups can fight each other for up to twenty minutes. I used to just let them slug it out with each other, but now I stop them after about ten minutes, just for safety's sake. I once lost a three-month-old pup when its littermate grabbed it by the throat and held on till he'd severed an artery. So now I watch their fighting more closely and break it up when it starts to get too frantic.*

"*By the age of four months, a game-bred dog from a good line can kill a cat. So that's when we start its real training.*"

All of this makes it quite clear that aggression and fighting tendencies can be bred into a dog. It also appears that the particular style of attack and the specific behaviors used to express aggression can be controlled genetically, as in the case of the Japanese Tosa.

Formal dog-fighting contests began in Japan late in the twelfth century. Historical documents indicate that, in the early 1300s, Hojo Takatoki, the fourteenth shikken (shogun's regent) of the Kamakura shogunate, was so obsessed with dog fighting that he allowed his samurai to pay their taxes with dogs. Dog fighting was encouraged because it was considered a way for the Samurai to retain their aggressive spirit even during times of peace.

Two types of dog fighting emerged that are still legal in Japan (except in Tokyo). The first is not much different from that in Europe and the Americas. This familiar form of dog fighting was popularized in the Akita Prefecture, which is the origin of the Akita breed. The Akita is a spitz-derived, or northern, breed of dog, with a curved tail carried high over the back, sharp face, and pricked ears. Akitas thus look like a shorter-haired version of the dogs that people commonly call "huskies." It is a big dog, around 110 pounds (50 kilograms) and stands about 28 inches at the shoulder (70 centimeters). It fights much like the bull terriers, only it is considerably less likely to lock its jaw on its opponent.

Another completely different form of dog fighting is modeled after the human sport of sumo wrestling. Sumo is an unusual form of wrestling in which large, well-muscled men try to stay on their feet, not allowing their opponent to pin them to the floor or drive them out of the ring. In the Tosa province (present-day Kochi prefecture), the Japanese Tosa (or Tosa Inu) dog was developed to compete in the canine equivalent of sumo wrestling. It is similar to the Akita in size and weight, but has a square blunt face, lopped ears, and ropelike tail (similar to Labrador retrievers), which all suggest it was derived from some kind of mastiff. This kind of dog fighting is called *Token,* and is the most popular type of dog fighting in Japan.

The fight is carried out according to the rules of sumo wrestling, and the winner is the dog that presses its opponent to the ground with its body, knocks it off its feet, or simply holds it to the ground. Dogs can also lose the fight if they bark, yelp, or lose the will to fight. Biting and growling dogs are disqualified and barred from further competition. Thus, the Tosa is really a "wrestling dog."

Champion dogs are called *yokozuna*, just as in sumo wrestling. Winning Tosas receive valuable, beautifully decorated cloth aprons topped with a braided collar of white hemp. This form of dog fighting does not have strong links to gambling in Japan.

Tosas' aggressive personality and behaviors are controlled by selective breeding. Tosas have been bred to wrestle—a form of aggressive expression that goes against the dog's natural instincts because it violates most of the normal fighting techniques used by dogs. Other dogs will bump, lean, rear up, and push down to show dominance, but once they become aroused they will use their teeth. Tosas are bred to be aggressive and dominant, but to inhibit using their teeth.

Kennel Clubs and Warnings

How successful has the breeding of fighting dogs been in altering the personality of some breeds? When the pit bull terrier was imported into the United States, the American Kennel Club refused to register them, arguing that they wanted no association with the sport of dog fighting or with the aggressive dogs bred for that activity. In response to this exclusion, Chauncy Bennet formed the United Kennel Club in 1898, initially as a breed registry designed solely for the documentation and acceptance of pit bull terriers. Bennet added "American" to the breed's name and initially dropped "pit" from the breed label but people who were using the dogs for fighting objected. They wanted the designation "pit" put back into the name, which thus resulted in the breed name becoming the American pit bull terrier. Bennet did not oppose this change since his aim was to create an organization that would represent the breed based upon each dog's performance. Initially, for a pit bull to be accepted into the United Kennel Club it had to have won three fights. This requirement was later dropped.

The American Kennel Club did ultimately decide to register pit bulls in 1936, but, perhaps to avoid the appearance of backtracking on their early stance, they did so under a different name. Now the pit bull was called the Staffordshire bull terrier. The name would be changed to the American Staffordshire bull terrier in 1972. This was done to clear the way for registration in 1974 of the British version of the pit bull, which could assume the now-vacant name of Staffordshire bull terrier.

Are pit bulls really more aggressive? The United Kennel Club now registers other breeds, although it remains one of the only kennel clubs that registers the American pit bull terrier (APBT). All kennel clubs tend to describe dog breeds in the most favorable light, usually using information provided to the national organization by the breed club. Breed clubs, since they mostly include breeders who want to popularize their breeds and sell their puppies, often go out of their way to find words that present their dogs in a positive, laudatory manner. Yet the United Kennel Club still puts a cautionary statement in the description of this breed on their website, saying, "Because most APBTs exhibit some level of dog aggression and because of its powerful physique, the APBT requires an owner who will carefully socialize and obedience train the dog."

Most breeders of pit bull–type fighting dogs protest that although their breeds might be aggressive toward dogs, they are not aggressive toward humans. They argue that even if actually used for fighting, these dogs have to be handled by the humans who feed them and care for them, so aggressiveness toward people is not permitted. Yet this is contradicted by those who train such dogs for fighting, such as Dave, who made it a point to mention that he and all of the other trainers he knew and worked with have all sustained bites from their dogs at one time or another. The breeders of pit bulls might argue that such bites simply occur because of the nature of the training, which enrages the dogs beyond normal limits. Alternatively, they excuse the fact that these dogs bite people as merely an unfortunate side effect of the fear and rage that is provoked during an actual dogfight where the dog feels that its life is being threatened (which it is). Unfortunately, the data contradicts this. Hard evidence shows that these breeds of dogs, which humans have specifically bred to be aggressive enough to be willing to fight and kill other dogs, are also significantly more aggressive toward people as well.

Dogs That Bite People

We need two pieces of data to determine whether pit bulls actually do have aggressive tendencies that are easily turned on humans. First, we need to know the number of pit bulls in the canine population and, second, we need information on the number of severe biting incidents. There are no accurate statistics that would give us the exact number of dogs of each breed in the United States. I did try to get information on the number of American pit bull terriers registered with the United Kennel Club. Despite the fact that my request for this information indicated that it would be used for research purposes, I was surprised to be told by Cindy Stickly of their registration department that "Unfortunately, these numbers are not for public knowledge." This is an odd statement given that virtually every other major national kennel club around the world routinely publishes information about the number of dogs in each of the breeds that they register. It made me wonder what the United Kennel Club was trying to hide. Fortunately, there is an alternative way to get an estimate of the number of pit bulls in the population.

Around one-third to one-half of all dogs in the United States is registered by the American Kennel Club, and this proportion is much higher if we consider only purebred dogs. No evidence suggests that some breeds are more or less likely to be registered than others. This means that to estimate the number of pit bulls, at least in percentages we can look at the number of American Staffordshire bull terriers, Staffordshire bull terriers, and bull terriers registered with the American Kennel Club. Of the 958,272 dogs registered in 2004, the combined total of these three breeds accounted for only one-half of 1 percent of all registered dogs.

Next, we need the number of dog bites. As mentioned earlier, in North America there is no national registry that tabulates statistics of all cases of dog bites. However there is one type of dog bite incident that must be reported—namely, those that result in the death of the victim. These dog bite deaths have been collected by researchers from the U.S. National Center for Injury Prevention and Control, who tabulated the results for the twenty years from 1979 through 1998. They found 238 dog-bite-related deaths during that period, in which the dog's breed was recorded. Based on the fact that pit bull–type dogs account for one-half of 1 percent of dogs we would expect only one or two pit bull–related fatalities if these fighting dog breeds were no more aggressive toward humans than other breeds. However, purebred pit bulls accounted for 98 fatal dog-bite attacks, which is 41.2 percent of all dog-related fatalities. If one includes the pit bull crossbred dogs, this number rises to a remarkable 49.6 percent of humans killed by dog attacks. This is an astonishing statistic, suggesting that pit bulls account for 80 to 100 times the number we would expect based on their population percentages.

While we are looking at these numbers it is interesting to look at another of the fighting breeds, that is, the Akita. They make up 0.3 percent of all dogs registered, but 1.7 percent of fatal dog attacks. This means that Akitas are more than 5 times more likely to be involved in fatal dog attacks than their numbers.

Judging by these numbers, breeders who have been trying to create aggressive fighting dogs through selective breeding have succeeded. Unfortunately, these dogs are not only dog aggressive, but under appropriate circumstances, they are aggressive to humans and such aggression can be fatal.

Converting Devils to Angels

The history of the fighting dogs has one more lesson for us about the creation of aggressive personalities. We can demonstrate that the genetic boulevard that leads to aggression in a dog breed is a two-way street that can also make a breed less aggressive if appropriate selective breeding techniques are applied.

The American Kennel Club recognized the bull terrier in 1885, but it was acknowledged as different from the pit bulls. I included its numbers along with those of the Staffordshires and American Staffordshires simply because those who report dog bites, including the various health and legal authorities, tend to classify it as a pit bull in their reports. There is some validity in this, since, as its name suggests, the bull terrier had started out as a pit-fighting dog that was used in bull baiting and was just as game and aggressive a dog as any of the other pit fighters. With the banning of pit sports in 1835, some breeders of the various fighting dogs decided to continue to breed game dogs, with the stated aim of producing compact guard dogs for homes and businesses, however with the tacit recognition that many of these could ultimately be used in illegal dogfights. An alternate path of action that was open to them was to change the personalities of these fighting dogs so that they would make more desirable companions and family dogs.

This is where James Hinks, of Birmingham, England, enters the picture. He decided that something new could be done with his fighting dogs. He believed that one breed (then called a *pit and bull terrier*) could be shaped to provide the gentlemen of the day with a elegant-looking companion, one that was calmer and more reliable than the dogs bred for fighting but that could still serve the function of a guard dog. His first addition to the genetic mix was the White English terrier, since he felt that white was a color that would be more attractive and distinctive, and more acceptable for the upper class, wealthy buyers whom he hoped to attract. For a while the breed was known as the "White Cavalier," however today few bull terriers are completely white, and color combinations as well as brindles are accepted. Some of the loss of pure white color was due to deliberate manipulations by later breeders who wanted a dog with a particular look.

Hinks recognized that the most important difference between his breed and others would not be their distinctive head shape and colors, but rather their personalities. To this end, like a chef trying to create an original dish, he mixed several other ingredients into the genetic soup that made up his breed. Specific breedings added characteristics of the Dalmatian, greyhound, the Spanish pointer, foxhound, and the borzoi. Physically, this gave the bull terrier a more elegant line and a somewhat more graceful bearing, since its legs were no longer bowed. More important, Hinks's mixture of breeds tended to offset some of the selective breeding for aggressiveness that characterized a fighting breed. Soon the breeders of bull terriers were advertising that their dogs were ideal family dogs with the potential to be a friend, a nanny, and a bodyguard at the same time. Today, although the bull terrier can be provoked to fight, it is generally no different in its level of aggression than any other terrier.

This selective breeding program has been generally successful. It is interesting to compare the United Kennel Club's inclusion of a warning about dog aggressiveness in their description of the American pit bull terrier to the American Kennel Club's description of the bull terrier which reads, in part, "The Bull Terrier must be strongly built, muscular, symmetrical and active, with a keen determined and intelligent expression, full of fire but of sweet disposition and amenable to discipline." The intent of the breeders of bull terriers can also be inferred from kennel names, which include Heavenly Kennels and Heartland Kennels, in contrast to Combat Zone Kennels or Armageddon Kennels, which produce American pit bull terriers.

The ability to breed aggressive tendencies *out* of a dog breed in the same way that fighting dog breeders selectively bred dogs to accentuate their aggressive personalities is important. It means that certain dog breeds, or certain bloodlines within these dog breeds, that now constitute a potential danger to other dogs and humans because of their aggressiveness can be remediated. Today's English mastiffs were once war dogs used to savage lightly armored enemy troops. In modern times, they have become "gentle giants" who are calm and unaggressive.

The Unflappable Bulldog

Perhaps the best example of breeding to remove aggressive tendencies is the bulldog, which was originally designed to be the perfect fighter when bull baiting was legal. It was shaped physically and mentally specifically to fight bulls. To that end, it was bred so that its shoulders are widely spaced and turned outwards, which allows it to crouch low to the ground so that it can dodge the bull's horns when it charges. The dog's head and forequarters are very well developed, so that as the bull charges by, the dog is able to spring to the bull's ears, nose, or throat. The bulldog's hindquarters are not as well developed as the forequarters, which means that once the dog has locked on to the bull's head it can be vigorously shaken without suffering any spinal injuries. The bulldog's rib cage is made of heavy bones and is well padded, so that if he does manage to latch on to the bull's nose (the perfect target to attack in order to bring the bull to its knees), and the animal tries to shake it off by slamming it into the ground, the dog can take the brunt of the blow. The shape of the bulldog's head allows a powerful bite. This bite is uneven, since the lower jawbone, or mandible, is longer than the upper jawbone, which allows him to hang on to its foe with a surprising tenaciousness, while grinding away to shred through skin and muscle. The bite is effectively locked, and the bull will not be released even if the dog is knocked out during the fight. The bulldog also has a short snout and the openings of its nostrils flare upward to allow it to breathe even after it has latched its jaws tightly on the bull.

The personality of the bulldog also had to fit its task, and it had to be easily provoked to violent, aggressive action. It had to be a tenacious fighter that would ignore pain and severe injuries, and would not break off an attack until its opponent was down, immobilized, or dead. These are all personality characteristics that it shared with the pit bull terriers. With the end of legal bull baiting and other blood sports, however, bulldog breeders chose a different course than the pit bull terrier breeders. Their aim was simply to produce a dog that looked as ferocious as its history implied, but that would be a good, calm companion and family dog. Quite specifically, it was bred for gentlemen who would be expected to spend time at their clubs, pubs, or gaming tables, or else sitting long hours at a desk or in a chair reading. This required a particularly placid dog who would not be easily stirred emotionally, and of course, displayed a minimum of aggression. Thus, the selective breeding pressure on the bulldog was the exact opposite of that on the pit bull terrier breeds and much like what happened to the bull terrier. The modern bulldog has been bred so that its former combative urges are no longer there. The contemporary bulldog doesn't see any sense in attacking a bull or a bear, and only in the rarest of situations can it be stirred to attack a human being.

Statistics bear out the success of bulldog breeders in changing their aggressive gladiators to ambling, placid family dogs. Using the same databases that we used before, we find that bulldogs constitute 2.0 percent of the registered dog breeds. However, this breed accounts for only 0.8 percent of fatal dog bites. This means that, numerically speaking, bulldogs are only around half as likely to bite as would be expected on the basis of their population numbers. This makes them a remarkably unaggressive breed. It also is a clear demonstration of how the aggressive component of a dog breed's personality can be reduced, as well as increased, through selective breeding practices.

Genetic manipulation also explains certain regional differences in dog breeds. For instance, Doberman pinschers and Rottweilers bred in North America tend to be somewhat calmer and less likely to initiate aggressive action than dogs of these same breeds bred in Europe and Latin America. North American breeders deliberately tried to tone down the aggressiveness in these breeds. In contrast, some European and Latin American breeders seem to prize and select for what they call "temperamental fire," which is really a willingness to display aggressive tendencies.

The degree of control that can be exerted on the personality of dogs, particularly to increase or decrease aggressiveness, is so great that when we encounter high percentages of aggressive dogs in any particular breed, the blame should be squarely laid on their breeders. Although poor socialization, abuse, fight and guard training, provocation, and other factors will certainly play a role in any particular dog-bite case, the genetic underpinnings are often decisive. Genes alter the chemistry of the body, and levels of certain hormones and neurotransmitters have now been implicated as being associated with canine aggression. For example, Karen Overall, of the Center for Neurobiology and Behavior at the University of Pennsylvania, has pointed out that dogs with a history of aggressive behaviors normally have higher levels of stress hormones, such as cortisol, in their blood. They also differ from dogs that do not have aggressive personalities in their usual levels of the neurotransmitters serotonin, dopamine, and the monoamines.

What does this say about suggestions that certain dog breeds should be restricted or even banned? Doubtless there are some, even many, American pit bull terriers, American Staffordshire bull terriers, and Staffordshire bull terriers that have been well bred and well reared and have the potential to become good, safe family dogs and companions. I have personally met, interacted with, and even become quite fond of a number of such dogs from these breeds. However, it is also true that many lines of these dogs have been bred for over two centuries to fight and to kill as part of the "sport" of dog fighting. Unfortunately, many of the dogs that are sold as pets or show dogs still have a close genetic relationship to those "game-bred" dogs. Scientists think in terms of probabilities, and as a scientist I can assert that if the specific provoking conditions arise, the probability is high that aggression will be triggered in those lines and breeds of dogs that contain the genetic heritage of dogs selected for their fighting abilities in recent generations—even if that animal has lived a peaceful and uneventful life until that moment.

Should we then be banning specific breeds of dogs as potential threats to public safety? I hope that this is not necessary. Perhaps in the near future, if the data from researchers like Overall prove to be correct, we can develop a chemical profile based on hormonal and neurotransmitter levels which will isolate litters or lines of dogs that should be restricted. Perhaps the Canine Genome Project will soon be able to isolate specific markers associated with aggressive tendencies. For the time being, since we are talking about genetics and genetically transmitted behavior tendencies, we should be able to isolate lines of dogs that are sound versus those that are potentially dangerous through the use of multigenerational pedigrees. Since it is not unusual to have three or even four generations of dogs available for testing or examination within any pedigree or breeding line, one could put together a reasonable estimate of how safe an individual family line of dogs is, rather than basing our judgments on group averages across a breed. Something like this must be done before public fears cause politicians to act in a more heavy-handed manner that may result in the virtual extinction of certain "at risk" breeds.

Regardless of the political and social paths taken to deal with aggressive lines of dogs, the data based on the history of the fighting breeds proves that the personalities of our dogs are under our control. If we want to create superdogs, we can. If we want to create monsters, we can. The responsibility lies with those who breed, rear, and train these dogs.

CHAPTER TWELVE

Heroic Hounds

Perhaps it is a holdover from our childhood fantasies but in our intimate thoughts, there is one canine personality characteristic that virtually every dog owner believes in. At some basic emotional level, we all believe that the one special defining aspect of dogs is their desire to help us. We also trust that our dogs will have the intelligence to recognize when help is needed and the courage to place their own lives at risk to save the lives of their beloved human family. Perhaps it harkens back to a primitive human huddled near a small fire, fearfully looking into the darkness yet somehow reassured because a dog is resting quietly nearby. Somehow, we all seem to draw comfort from our belief that in a time of crisis our dogs will turn into heroes, saviors, rescuers, or faithful defenders, just like Lassie, Rin Tin Tin, Benji, and every other dog star we have seen depicted in the movies.

There seem to be frequent enough examples to justify such feelings. A typical case in point occurred in Vermont, involving an English setter named Bruce. His owner, Lewis Gardner Reynolds, set off one morning to hunt ducks on Silver Lake. Although a retriever would have been more suited to fetch the ducks from the water, Bruce had learned how to retrieve well enough to warrant his presence on the hunt. When the man and dog arrived at the lake, they saw that along the shore was about 20 feet (6 meters) of ice that extended out into the open water. Reynolds gathered some brush and reeds and made an impromptu blind near the far edge of the ice to hide in while he waited for the ducks to arrive.

The hunt seemed to be going quite normally. Several flights of mallards had gone by and Reynolds had shot two. Bruce had retrieved them from the water and was patiently standing a short distance behind his master until he was needed again. Unfortunately, the early winter ice was not very strong and began to give way under the man's weight. In the excitement of the hunt, Reynolds did not notice this until his boots were ankle deep in the cold water and it was too late to get back to shore. Cautiously turning toward the shore, he took a step, but the ice began to crack and he knew that if it broke he would sink into deep water. With his heavy boots and winter clothing, there was no doubt that he would be dragged to the bottom. Thinking quickly, Reynolds let himself fall forward, spreading his hands and feet out to distribute his weight as widely as possible over the fragile surface. He felt the slick of the bitterly cold water that covered the surface of the crumbling ice seeping through his clothes and chilling him as he tentatively tried to creep toward the shore. He moved slowly so as not to put too much strain on any part of the ice, but he made virtually no progress on the slippery surface. More vigorous movements only caused the ice to crack more.

Reynolds later described the situation. "*I knew that I was in a predicament and could imagine myself freezing to death when night came or drowning if the ice broke. The shore looked like it was miles away. But there was Bruce, crouching nearby and looking at me with an expression that was as nearly human as I've ever seen on an animal. Suddenly I thought of a possible way out of this mess. At least it was the only one that I could think of that seemed worth trying. I took off one of my winter gloves and tossed it to Bruce and gave him the command to 'Go!'*

"*He seemed to know what I needed and grabbed the glove. I remember seeing that streak of black and white as he went from the ice to the shore and into the woods.*"

Bruce was not in any danger of falling through the ice since his weight was much lighter than his master's and more evenly distributed. However, the slickness of the surface still meant that he could have offered little help to the man directly. When Reynolds had fallen on the ice and wriggled helplessly, it could not have meant anything intelligible to the dog, who might have wondered what kind of new game his master was trying to play. However, he most certainly would have recognized the distress in the man's facial expression and his voice when he threw the glove to him with the command "Go!"

It is unclear how well Bruce knew the region near the lake, but he appears to have run directly to the nearest farmhouse, which was about a half mile (1 kilometer) away. There he found the farmer, Benjamin Ross, out doing some evening chores. The dog then attracted his attention by shaking his head frantically with the glove still in his mouth. He then dropped the glove, crouched over it, and barked. When Ross stepped toward him, the dog grabbed the glove and started to move back in the direction he had come from. It just seemed to the farmer that the dog was trying to tell him that he wanted him to follow. He wondered to himself, "Perhaps someone is in trouble and needs help." He dropped his tools and followed the dog. Just as he was going, by some inspiration, he also grabbed a long length of rope.

Reynolds takes up the story from there. *"Bruce piloted him to where I was lying sprawled out and sinking down into the water. Ben made several efforts to toss the rope out to me, but it always fell short and I couldn't move to reach it. Bruce finally lost patience with all of this. He grabbed the rope and carefully carried it out to me. After that, it was an easy matter for Ben to drag me over the smooth ice to the shore. When I was finally on solid ground Bruce seemed to be frantic with joy and that dog didn't care who knew it!"*

When we hear a story such as this, we might have several emotions, such as admiration for the dog's intelligence and devotion, but seldom does the average person have a feeling of disbelief. Somehow we all know that "dogs are like that," "it is just part of their nature," and it confirms our belief that it is in the personality of any dog to become a hero if the need arises.

A Gathering of Heroes

Stories about dogs saving the lives of people have been told for centuries. Here I am not speaking about dogs that have been specifically trained for service, such as tracking dogs, protection dogs, police dogs, war dogs, or search-and-rescue dogs. Rather I am referring to pet dogs that act spontaneously to assist humans. Remarkably, though, when I searched the scientific literature I could not find any research that was designed to study this aspect of dog behavior. Thus, I was left with the task of gathering the data myself.

I tried to get a sense of the nature and frequency of dog heroism by electronically searching the archives of four national newspapers: the *New York Times, Wall Street Journal, Washington Post,* and the *Christian Science Monitor.* I knew that most stories of dogs rescuing people don't appear in such "serious" national publications, but in more local newspapers (if they make the news at all). Unfortunately, such smaller publications often don't have easily searched back issues. Over the years, however, I myself had accumulated a substantial collection of clippings from a variety of publications that described the activities of heroic dogs.

Since I was interested in dogs' natural behavior disposition to help people, I felt it necessary first to eliminate all of the stories involving trained service dogs. There were a great many of these, especially in more recent times. This excluded reports concerning guide dogs, police patrol dogs, and search-and-rescue dogs that have been deliberately trained to assist people. Even if the dog was actually acting heroically by doing something that it was not trained for as part of its jobs—such as a guide dog defending its blind owner from the attack of a mugger or a police patrol dog saving a child from drowning—there is a possibility that specifically training the dog to help in one set of tasks might have a wider influence on the dog's mind and predispose him to help in other situations. Next, I eliminated any duplicate reports and any stories that just appeared to be too incredible to be believed (such as "Dog Drives Sick Owner to the Hospital" or "Dog Senses Owner Is in Danger from a Distance of 6 Miles"). My target was to collect at least a thousand cases to allow some reasonable sort of statistical analysis. In the end I stopped at 1,006, although I could have added many more if time permitted. Still, I felt sure that I had enough information to get a general picture of the ways that dogs act heroically.

Sounding the Alarm

If you ask most people to name the situations in which dogs are most apt to act heroically and help save the lives of people, the vast majority seem to think of Lassie and Rin Tin Tin leaping toward the evil masked bandit or saving a young child by driving off a marauding wolf. The general belief seems to be that the dogs that most frequently turn out to be heroes are dogs that serve in the role of protectors of people, in the sense of being warriors or bodyguards. This is certainly reinforced by the fact that the most commonly encountered service dogs are guard dogs or police patrol dogs. I was surprised to learn that protecting people from assaults, threats, or attacks was not the most commonly reported situation in which dogs come to the assistance of people. Most frequently, dogs intervene to save lives by alerting people to danger, such as to fire, gas leaks, carbon monoxide, or the presence of a dangerous situation, such as a lurking snake or other hazard. This warning behavior accounted for approximately 35 percent of the reports.

A typical example of such warning behaviors occurred in Mansfield, Pennsylvania, involving a Rhodesian ridgeback named Fudge. Sharon Dennington and her partner were asleep when Fudge climbed onto her bed. Dennington later reported, "She was barking and she'd got my hair in her teeth. Then I noticed there was all this smoke coming through my bedroom, and she was really telling me there was something going wrong." Fortunately, the warning was early enough to allow Sharon and her partner to gather up her two children and get safely out of the house. As a result, Fudge saved four lives that night.

Another dog who provided a life-saving warning was Butch, an Alaskan malamute in Wausau, Wisconsin. Carbon monoxide was leaking from a faulty furnace. Butch ran to his mistress, Carla Martin, and by whimpering and whining, was able to awaken her. Unfortunately, the concentration of the gas was so great that she collapsed before she could call for help. Butch stayed by her, nudging and licking at her until she regained consciousness. Once she had regained her senses, she was able to cry out for her husband Steve, who was in the next room. Steve didn't know exactly what the problem was, but seeing his wife's condition he immediately summoned help. Meanwhile, Carla staggered her way to the bedroom of their ten-year-old son, Mitchell and helped him escape to the outside. Shortly afterward, an emergency service worker carried their sixteen-year-old daughter, Shana, who was unconscious, out to the front lawn. She had been sleeping in a bedroom on the lower level of the home where the concentration of the gas was the greatest.

Carbon monoxide is particularly deadly since it is invisible to the eye and odorless to humans, but apparently not to dogs. Steve summarized the situation quite simply when he said, "He knew something was wrong. He saved our lives."

A skeptic might look at such cases and conclude that there was no heroism or conscious decision to help involved in such reports. The skeptic's argument might suggest that these were merely instances where a dog detected a potentially threatening situation, became worried and anxious, and wanted to escape for its own safety. If this is true then it was just the dog's distraught emotional behaviors, evoked by its concern for its own well-being that unintentionally warned his family. Alternatively, one might credit the dog with more intelligence but still believe it was acting in its own interests by alerting his human companions. In this scenario, the skeptic suggests that perhaps the dog reasoned that the only way to escape the problem was to awaken family members, since they are the ones who normally let him out of the house and could thus provide him with an avenue of escape. The fact that lives are saved in such situations is thus interpreted as simply an accidental outcome of the behaviors of a frightened dog worried about its own salvation.

This analysis does not work, however, since there are many cases where a dog is outside the house or room where there is a gas leak or fire, and makes valiant efforts to get inside and go toward the problem—into danger—in order to warn his human companions. As an example, consider the case of King, a German shepherd and husky cross living in Granite Falls, Washington. It was Christmas night and the big dog was sleeping in the recreation room. A sliding door to the outside was always left open, so that King could go outside when he wanted, and the door that led to the rest of the house was closed to keep out the cold night air. Late that night the dog awakened and immediately was alerted to the smell of smoke seeping from under the door that led to the interior of the house.

If the dog's main concern were his own well-being it would have been easy for him simply to leave through the open sliding door and wait outside in complete safety. However, three of the people that this dog loved most, Fern and Howard Carlson and their daughter Pearl, were inside, where he apparently sensed that a fire was exposing them to danger. Instead of running out, King turned his back on safety and began tearing at the door that separated him from his family. Fortunately, the door was hollow-cored, and a gap split open which gave him something to grab on to. Using his teeth and claws, he ripped at the veneer, pulling off strips off the door. Despite the fact that he now had splinters embedded in his mouth, he continued biting and clawing in an effort to widen a hole and gain entrance to the house.

By the time King had created a substantial opening in the door, the room on the other side had begun to burn. Now in order to get to the Carlsons he would have to actually go through the flames. At any time, this dog could have simply turned and escaped through the open door to the outside. However, King ignored this option, returning to his efforts to rip at the plywood until he had a hole that was sufficiently wide. Once he thought that the opening was large enough to permit him to go through, he inserted his head in the gap and charged forward. The aperture was tight, and the powerful dog scraped the edges with such force that jagged pieces of the door tore at his neck.

King was now standing inside a small utility room which was engulfed with flames and filled with choking smoke. The fur on his head and shoulders was scorched and there were painful burns on the pads of his feet. Nonetheless, he raced down the hall to the room where sixteen-year-old Pearl slept. She was deeply asleep, and as he tried to nudge her awake with his nose, she shoved him away and rolled over. Now the dog increased his efforts and tugged vigorously at her nightclothes until she sat up. Although she was groggy and was about to scold King for awakening her, the acrid smell of burning walls and carpets let her know that something was wrong.

With King beside her, Pearl ran to the door of the bedroom where her parents, Fern and Howard, were sleeping. By the time Pearl shook her mother awake, the smoke had already begun to fill the room and Fern recognized the danger that they were in. Fern poked at Howard and shouted for him to get up and get out of the house, then she and Pearl went to the bedroom window. Since the house was only one story and the flames seemed to be everywhere, this was the most natural means of escape. Once they were safely on the outside, King stood by the window looking out at Pearl and Fern. They were standing on the lawn and calling for him to jump. Once again, with safety clearly in front of him he turned away and returned to the dangers of the burning interior.

What King understood was that Howard, who had a respiratory problem, had not gotten up from his bed. Ignoring the pain from his burns and the gash in his neck, the dog went back to Howard and repeated the same behaviors that he had used to rouse Pearl, nosing, pawing, and whimpering to get him to move. When his master awakened, however, he was weak and confused. He slowly got up, and then awkwardly started moving in the wrong direction, away from the window. He was trying to make his way down the hallway, which was already partly engulfed in flames. Standing in the doorway between the bedroom and the hall Howard seemed to be confused, but before he could move farther he simply crumpled to the floor and lay unmoving. King was with him when he fell, and immediately ran from him to the window and barked frantically. Pausing only a moment to look at Fern and then back to the interior, he then ran back to the inert man. He repeated this run to the window, again barking while looked back and forth from Fern to the prostrate form of Howard. Then he disappeared from sight as he raced back to his master. This time Fern thought that she understood that Howard needed help. The moment she climbed back in through the window, King led her to her husband.

Fern desperately worked to rouse Howard. He rose and put one arm around her while the other hand tightly gripped King's collar. With Howard supported in this way the group moved slowly. Nonetheless, the three of them managed to stagger back through the smoke and out the window to safety. Howard lay gasping on the ground with Fern kneeling next to him and all the while King hovered close in a protective manner. Pearl had run to the neighbors for help, and fire trucks and emergency vehicles converged on the rubble that had once been the Carlson home.

As the morning light dawned, it became clear what King had endured. His mouth was full of painful splinters, the fur running down his back had been singed off showing bare skin, and blood seeped from an ugly wound on his neck. The pads of his paws were badly blistered and it would be a year before they returned to normal sensitivity. He had endured all of this and refused to seek sanctuary for himself until his family was safe and secure. These were certainly not the acts of an anxious dog focused only on his own well-being, but rather a clear attempt to render assistance to the humans that were important to him.

Bringing Help

The second largest category of dogs saving lives accounted for 22 percent of the cases. This included those situations where dogs bring help to people in trouble by finding humans that might serve as rescuers and leading them to the person who is in danger. This was clearly the most important thing that Bruce did at the opening of this chapter, when he brought help to his master stranded on the ice. Bringing the rope to his master was an added bit of helpfulness. About half of the cases in which dogs brought help involved people who had suffered injuries or were experiencing physical difficulties from heart attacks or a diabetic incident. Another third of the cases involved bringing help to trapped individuals. The remainder of the cases in this category involve bringing rescuers to lost people. Obviously, some incidents involve multiple problems, such as when a person falls down a well, incurring injuries, and is unable to climb out. This could be counted either as being trapped or injured. I was somewhat arbitrary in my scoring of such cases, counting each depending upon the problem that seemed to be most severe and most immediately life threatening.

The pattern of most of these rescues is similar. Take the case of

Max Lovett of Alliston, Ontario, in Canada, and his bouncy, energetic Irish setter, Caleigh. Max usually took Caleigh out for a long walk each morning, but on this winter morning he was not feeling all that well. He knew that the spirited dog needed her exercise, but he had a backup plan for days when walks were not convenient. He drove Caleigh out to a nearby farm where he had permission to let the big red dog run off leash. Max stood in the field and watched Caleigh frolic in the snow but felt himself beginning to feel quite sick and short of breath. Thinking that he would be better off sitting in the car he turned toward it but had only gone a few steps before he collapsed. A few moments later he returned to consciousness only to find himself lying facedown in the snow and unable to move. Caleigh saw her master fall, immediately stopped playing, and returned to his side. Sensing something was wrong, she whimpered and nosed at the fallen man, then began to look around frantically.

Suddenly Caleigh caught sight of Mike Raworth, the farmer who owned the land. He was outside his house shoveling snow from the steps to his front door. Caleigh dashed across the field. When she reached Mike she barked and dashed in the direction where Max lay. When the man didn't respond, she returned to him, barked again, then dashed in the direction of her master. Mike leaned on his shovel and looked puzzled at this behavior. Then Caleigh again rushed toward him, this time jumping up and spinning around a couple of times before running toward the field. Finally, the farmer realized that the setter was trying to tell him something, or more precisely, to show him something, so he followed her across the field. There he found Max stretched out in the snow. When he tried to rouse him and got no response other than some twitching movements, Mike ran back to the house and called an ambulance. He returned a few minutes later with a blanket, wrapping it around Max, while Caleigh pushed herself close to the fallen man as if she was also trying to keep him warm.

When the paramedics arrived, Max was barely conscious. They quickly concluded that he had suffered a heart attack. In addition, his body temperature had plummeted to a level where hypothermia could be dangerous. One of the paramedics commented "Good thing we got here when we did. Much longer out here in the cold and he'd be dead." Max was quickly transferred to a hospital, and because of Caleigh's actions, he is alive today and can still walk his handsome red dog.

Another example involves Sophie le Roux who lives in Bloemfontein, South Africa. In her seventies, Sophie was still active and often traveled to Queensland in her car, always taking her German shepherd dog, Leila, for company.

On this particular evening the mist was quite thick and Sophie missed a turn in the unlighted road. Her car plunged down a steep embankment and landed on its side. The car's frame crumpled in a way that pinned Sophie in place and made it impossible to release the seat belt or open the doors. Leila had been thrown from the automobile when the windows smashed as the car rolled down the sharp slope.

"I tried to get out but couldn't move," Sophie said. "Then I heard Leila barking. She was running around the car and making a lot of noise. I thought that that was a good thing since no one would see me down there especially with all of that mist, but maybe they would be attracted by the sound of her barking. A few minutes later Leila disappeared. I think that I panicked then, since without someone to show the way to where I was there was simply no way that I would be found alive."

Leila, however, had not simply abandoned Sophie. She had clambered up the slope to the road and then retraced the path that the car had taken. Some distance down the highway she found some road workers. She ran toward them barking, at first raising their fears that she might be threatening or aggressive, but then she turned and ran back in the direction of the car looking back over her shoulder to see if they were following. When they didn't respond, she repeated the pattern, running toward them, then turning away to lead them in the direction of her mistress. Ultimately, one of the workers became convinced that she was trying to get them to follow her.

When the men got to the place where the automobile had slid off the road the car was not visible because the heavy mist had settled into the lower elevations. Leila, however, charged down the slope and the men could hear her barking in the mist. It was then that one of them noticed that ground was ripped up as if a heavy object had fallen over and began to understand what had happened. The workers carefully made their way down the embankment and with some difficulty managed to free Sophie. Her first words to both her canine and human rescuers were that as soon as she was able to do so she would bake them a cake and treat them all to really big pieces.

Leila and Caleigh both used the most typical means of getting rescuers to come to the aid of people in need—barking, whining, or whimpering in an excited manner, and then running in the direction where the person requiring assistance was located. This seems to be an instinctive means of communication in dogs. This natural behavioral tendency is the foundation for training hearing assistance dogs, whose instinct is modified so that they alert the deaf person by pawing, jumping, nudging, or other attention-getting behaviors and then running in the direction of particular sounds (which could be a doorbell, tea kettle, telephone, or other significant sound-producing object). Once the dog has the person's attention, it runs back and forth between the object of interest and the person until the human ultimately follows and discovers the source of the sound.

There are other ways of bringing rescuers to a scene. In 1907, in Oxford, Pennsylvania, a collie used a unique method. A farmer named William Beattie and his family were saved from being burned to death by some very clever actions taken by their dog. A fire started in one of the back rooms of their house late at night and the dogs on the farm began barking. Unfortunately, no one paid attention to them. One of these dogs, their collie, however, came up with another way to sound the alert and bring help. He had been trained to ring a bell to signal the men working in the field that it was time to come in. When his barks did not bring people to the site of the problem, the inventive dog raced to the bell, grabbed its cord in his mouth, and began to tug as hard as possible.

The loud tones of the bell carried quite a distance and roused Beattie and his family. It was also unusual for the bell to ring in the middle of the night and so this caught the attention of some nearby neighbors who came out of their home to see what was happening. When they caught sight of the flames, they ran to the rescue. Finally fully awakened, the father found the entire house burning. He immediately raced to the room where his two young boys were sleeping.

Both were already partially overcome with smoke, but, with the assistance of the neighbors, both were carried to safety.

It appears that the dog, whose barks had not been answered, used a learned association to bring help. He knew that the ringing of the bell usually resulted in people coming to his location and recognizing that humans were needed to help his family, he used this learned response to bring rescuers to his loved ones.

Pulling, Guiding, and Nudging People to Safety

Dogs frequently sound the alarm or alert to danger, and sometimes bring help, but it also appears that there are times when a dog will physically rescue a person in peril. Such instances amount to nearly 20 percent of the reported cases of canine heroism. The largest share of these involve saving a person from drowning or pulling them out of the water when they have broken through the ice. One historically significant example involves Maurice Maeterlinck (1862–1949), a Belgian author and one of the founders of the Symbolist style, which focused more on mood and emotion rather than straight storyline. Maeterlinck's work was so influential that in 1911 he was awarded the Nobel Prize in literature. One day, when he was still a young child, he was out playing near a river. While running to try to catch a butterfly he stumbled on a tree root, tumbled down the bank, hit his head, and rolled into the water. Barely conscious and unable to swim, he was beginning to be carried into deeper water, while his younger sister could only look on and scream in horror. Suddenly, the family dog, a great black-and-white Newfoundland, ran past her and splashed into the water. The big dog grabbed the child by his shirt and pulled him out of the water and up the bank. While he stood over the boy, licking his face, the girl ran for help. Had it not been for that dog some sixty major works of drama and literature would not have been written. Included in that body of writing which might never have come into existence is one of the most moving works ever written about dogs, namely Maeterlinck's *Our Friend the Dog.* It is a philosophical yet personal speculation on the lives of dogs and their relationship to humans.

That little book was inspired by the death of a young bulldog that he owned, not by the Newfoundland who had saved his life.

Many water rescues are quite dramatic, as in the case of Rob and Laurie Roberts of Glenwood Springs, Colorado. They had decided to shoot the rapids of the Colorado River in a small rubber boat and were accompanied by their Labrador retriever, Bo. Unexpectedly, the river narrowed to a boulder-choked channel of frothy water. The speed of the river increased and in moments their little craft was spinning out of control. Suddenly the raft hit a massive rock, and was vaulted high in the air, flipping over as it fell back into the raging torrent. Rob was thrown sideways into a clearer section of the channel, swept downstream a distance, and finally managed to crawl ashore. Laurie and Bo, however, were pinned under the raft. Rob looked on helplessly, unable to fight the strong rush of water in order to get back upstream to the raft.

The woman and the dog found themselves in a struggle for their lives. The pressure of the current pushed the flexible boat against them, smothering them while at the same time scraping their bodies against bare rocks as it shoved them against the canyon wall. Somehow, Bo managed to pull himself free and surfaced in surging water. However, Laurie, despite all of the straining that she did against the tough rubber, remained pinned in place, as though a giant foot was standing on her.

Now Bo showed his heroism. Although he was gasping for air and straining against the cold current to hold his position, he refused to swim to the safety of the shore, but instead returned to the capsized boat. Next, he dived beneath the surface, struggling against the pressure of the water that was apparently determined to push him downstream. Finally, he made his way back to the side of his drowning mistress.

The air under the raft was nearly gone, and Laurie was completely exhausted from her struggles, when the dark head of the dog popped up beside her. Bo grabbed at the first thing that appeared available to grip, which resulted in him sinking his teeth into a mass of Laurie's long hair. She barely had time to shriek in pain and panic before the dog dragged her down, away from the rubber raft that seemed to be trying to suck them back into its hold. After a few strokes, he turned sharply toward slower-moving water and then headed up to the surface. Moments later they broke the surface.

Although breathing was no longer a problem, the river continued to bounce them violently into the rocks. Laurie's energy was beginning to fail and the possibility of still drowning loomed close. Then Bo swam to Laurie and pressed his body against hers and waited until she grabbed hold of him. Laurie would later have difficulty remembering whether she grabbed his collar or his tail, but wherever she gripped him, her hold was strong enough to allow the dog to pull her to the safety of the gravel-strewn shore. When Rob finally reached them, they were both still coughing up water and shivering from exhaustion.

One might argue that the basic behaviors involved in this water rescue were instinctive. Most dogs will retrieve, and a Labrador retriever has been specifically bred to drag things from the water. Yet much more is clearly involved here since Labrador retrievers have not been bred to retrieve anything as large as an adult human being, nor to put their lives at risk while doing so. As in many other instances where dogs have tried to help, Bo had chosen not to seek his own safety first, even though opportunities to escape danger existed. Instead, his first priority seemed to be his desire to help a member of his human family.

While saving people from drowning is one way that dogs directly intervene, another situation appears in numerous news reports. Ever since man developed a love of fast travel, dogs have been around to push people out of the path of speeding vehicles. Such reports of dogs saving people published in the mid-1800s usually involved carriages, horses, and trains, while later reports are mostly about cars and trucks. Regardless of the nature of the vehicles involved, all of these reports seem remarkably similar. Let me give just one example.

Top was a handsome Great Dane living in Los Angeles with his owner, the German-born actor Axel Patzwaldt. Although big, he was a gentle dog who was popular with the children who lived in Patzwaldt's apartment complex. On this warm April day the eleven-year-old daughter of a neighbor had been begging Patzwaldt to be allowed to walk Top. She was so persistent that he finally agreed. He clipped a leash onto Top's collar and watched them walk down the sidewalk. At the end of the block the girl was still quite excited about her adventure and was looking at Top rather than checking the street before she stepped off the curb. Unfortunately, she was directly in the path of a quickly moving delivery truck. It was too late for the truck to stop. Top sensed the danger but only had time for one deep warning bark. In a single sinuous move, Top spun himself in front of the girl, pushing her with his shoulder with enough force to cause her to topple back onto the sidewalk in time to escape unhurt, before the truck rolled over the exact place where they had been standing. Unfortunately, Top had not had time to fully remove himself from the path of danger, and the heavy bumper of the truck hit his rear, sending him flying into the air. He crashed to the street in a crumpled heap and had time for a single moan of pain before losing consciousness. Top's right hip and hind leg had been broken and he would walk with a limp for the rest of his life, but the little girl was still alive because of the big dog's direct physical intervention.

The remaining rescues in this category involve dogs that guide, nudge, or pull people to safety. This often involves leading someone who is lost to help, or dragging or guiding someone out of a fire. An interesting variation on this theme occurred in Chinnakalapet, India, one of the places hit by the tsunami of 2004 which killed more than 150,000 people. The great wave was caused by an earthquake, and many of the Southeast Asian towns and villages that nestle near the coast had virtually no warning of impending disaster. That morning in December seemed quite normal, with sunny skies and a cool breeze. R. Ramakrishnan, a fisherman, had just returned from his morning's work with a boat full of fish. His wife, Sangeeta, was preparing food and their three children were playing nearby. Sangeeta heard a strange noise coming from the direction of the sea, but a view of the ocean from their home is blocked by a two-storey community center. Ramakrishnan decided to investigate and climbed to the roof of the center from where he saw the giant wall of waves approaching. He immediately ran to the edge of the roof and shouted down to Sangeeta, "Tidal wave—run for the high ground!"

Sangeeta quickly grabbed her two youngest sons, and yelled to Dinakaran, her firstborn son, "Run!" It was a sensible thing to do, since he was the oldest and at seven years of age stood a good chance of being able to outrun the oncoming tsunami. Now frightened, Dinakaran didn't tag along behind her as expected. Instead, he ran to the safest place he knew, their small family hut, located a mere 45 yards (around 40 meters) from the seashore.

Sangeeta made it to the high ground across the main road, but when she realized that Dinakaran had not followed her but instead had run home she collapsed into tears, screaming over the loss of her eldest son. "I had heard from others that the wall of my house had collapsed, I felt sure that my child had died," she later said.

Fortunately, a yellow dog, with a sharp pointed face and erect pricked ears, named Selvakumar, intervened. Dinakaran is clearly the dog's favorite child and escorts him to and from school each day. The rest of the time he is happy just spending the day playing with Dinakaran's brothers or hanging around the family and begging for food. At night he finds a way to sleep with the children, even though their parents have tried to keep him out many times. Selvakumar's name is special. When he was a puppy, he was given to Sangeeta by her brother-in-law as a gift following the birth of her second son. Unfortunately, her brother-in-law died in an accident, not too long after, and they changed the dog's name to his, to honor his memory.

Water was already beginning to rise near the hut when Selvakumar ran in after Dinakaran. The boy was confused and frightened. He simply could not think of what to do next when he was now confronted by his dog who was barking loudly at him. Selvakumar next tried nipping at the boy to move him out of the house. The dog then ran to the door and barked, and then back to the boy, giving the usual canine signal meaning "Follow me!" When the boy showed no evidence of understanding the dog actively attempted to nudge the child out of the door. Selvakumar clearly seemed to be running out of ways of communicating with the child who was still confused and not responding. Finally, according to Dinakaran, "That dog grabbed me by the collar of my shirt. He dragged me out." Now outside, the boy saw the last of the villagers running toward the hill that the main road skirted. The yellow dog barked and ran toward them, coming back to nudge him again. At long last the boy understood and began to race for safety with the dog scampering a few paces ahead of him.

Sangeeta says, "He saved my child's life. Perhaps he is guided by some special spirit—probably that of my brother-in-law, since he was blessed with his name."

Canine Warriors and Guards

In the movies, heroic dogs are most often called upon to do battle. Usually this requires them to protect their master physically from bad men with guns or from marauding grizzly bears or mountain lions. While this is the most common task of the Hollywood film dog, it ranks fourth in terms of actual rescues reported in the newspapers and accounted for just under one in five (18 percent) of the heroic acts reported. A typical story comes from Marylyn Johnson of Miami.

"My husband's work often takes him out of town and he decided that I should have a dog to protect me. He got me a handsome Doberman pinscher that I named Sherlock.

"Sherlock is not at all like what people think Dobermans are. He's laid back, likes people, and likes to play. I knew that he would be a complete failure as a guard, but he certainly looked tough enough to discourage anyone who had evil intentions.

"One night, when Steve was away I was sleeping in our second-floor bedroom and Sherlock was sleeping next to my bed. Suddenly he started growling, which is something he never does. I was sort of waking up but was still foggy. I told him to keep quiet, but noticed that he was looking at the sliding glass door to the outside balcony. When I looked in that direction I saw a man coming in from the outside. He had something in his hand.

"Suddenly Sherlock leapt directly at his face. The man yelled and started swinging his hand at my dog. Then he backed up and tried to close the door between himself and Sherlock, but Sherlock jumped at his chest before he could, and knocked him over the edge of the balcony. I called the police and they found him right away. He had broken his leg in the fall, and he was covered all over with blood. It turned out that it was Sherlock's blood, since he had managed to stab him three times. Sherlock's wounds took a while to heal, but he is a hero. The man's DNA proved that he had raped and beaten at least one other woman in the neighborhood only a few weeks earlier.

"I've never heard Sherlock growl again since then, but now I'm sure that he really is the guard dog that I want. I now know that he will be a

sweetheart most of the time, but when the chips are down he's willing to risk his life for me."

Sometimes even quite small dogs acting defensively can deter a human attack without requiring any physical intervention. For example, Jeff Bellis, the manager of the Albion Pub in Manchester, England, was closing down just after midnight when there was a knock at the door.

According to Bellis, "When I heard the door I thought it was one of the staff coming back for something. But when I opened it, there was this man with a gun. He was wearing a black balaclava over his face and was trying to push his way in. My girlfriend Rena was in the pub at the time so I started shouting and tried to close the door, but he kept pushing. I think it was because I was shouting that Foxy realized something was wrong. He started barking and the hair on his back went up. The guy just ran off."

Foxy is only a 10-inch-tall (25 centimeter) Jack Russell terrier. Nonetheless, in his rage to protect his master, he produced a truly threatening, larger-than-life bark.

"It wasn't just any old bark; Foxy was really standing his ground, and I think that's what scared the bloke off. I never thought a dog that size would scare someone away."

Foxy had just proven the truth of one of Dwight D. Eisenhower's favorite maxims. The former U.S. president and general liked to say, "What counts is not the size of the dog in the fight; it's the size of the fight in the dog."

This maxim has been proven many times, when small dogs have confronted men or animals that were considerably larger and stronger then they were to try to help people that they cared about. One such case involved Jakob Jelenic, aged sixty-five, who found himself stranded in a remote section of central Australia. While trying to work his way back to civilization he encountered a wild water buffalo. Adult males are easily angered and with horns measuring 13 feet (4 meters) from tip to tip along the outside curve across the forehead, they are so formidable that even a tiger will shy away from attacking them. Recognizing that his life was in peril, Jelenic tried to get away, but for some reason the big animal was enraged and kept threatening and charging at him. The pursuit extended over a mile (two kilometers) and only ended when the dangerous animal was eventually driven off by Jelenic's dog. The dog nipped at the buffalo's legs and harried it until the large beast finally found the situation too uncomfortable and gave up the chase.

To understand how heroic his dog was, you must understand that the water buffalo stands 5 to 6 feet tall (1.5 to 1.8 meters) at the shoulder, and weighs over 2,000 pounds (around 1,000 kilograms). Compare this to Jelenic's fox terrier that stands 15 inches (39 centimeters) at the shoulder and weighs 18 pounds (8 kilograms). With that much difference in stature, it takes a dog with a lot of fight in him to continue to attack rather than running for cover.

Clearly, dogs do engage in helpful, heroic acts that often save people's lives. They do this without training and often at great risk to themselves. The tasks that they set for themselves, in the order of their frequency in published reports, are: alerting humans to danger; summoning help to humans who are injured, trapped, or in peril; physically dragging or guiding humans to safety; and serving as warriors and protectors. Understanding that to be the case, our next task is to find out why dogs perform such selfless acts, and if there are any breeds that are most likely to rush to the aid of their human companions.

CHAPTER THIRTEEN

Why Do Dogs Help?

Why do dogs spontaneously help people? We have already seen that dogs often risk their own lives and safety to help human beings. However, a strict traditional view of how evolution works might suggest that they should not.

The theory of evolution was based on a concept of "the survival of the fittest" where fitness is measured by how many offspring an animal will produce. A simple example is that green insects that feed on green leaves and mottled-gray bark-feeders that sit on mottled-gray tree bark are more difficult for birds and other predators to see and capture than are insects of more visible colors. This means that these camouflaged animals have an advantage and are more likely to survive and reproduce, which means that they are "fit" for their environment.

Charles Darwin envisioned the fitness principle working asserting itself, not only through physical characteristics, but also with behavioral characteristics that can be coded in the genes. Any behavioral predisposition that might allow animals to live longer and produce more offspring should then be passed on as part of the genetic heritage of the generations that follow. In this way, particular behaviors become dominant in any species.

Altruism in Animals

When I described the behaviors of heroic dogs, I suggested that most people believe that such heroic or helping tendencies are part of the personality of dogs. One might even speculate on the combination of personality traits that might lead to heroic behavior, such as above-average sociability and energy level and below-average fearfulness or some such recipe. When we are talking about human behaviors, psychologists view heroism as an extreme example of *altruism*, which is defined as an unselfish concern for the welfare of others. Some individuals appear to be altruistic most of the time, while there are others that are not, so in some respects altruism can be viewed as a behavioral predisposition, much like a personality trait.

The problem is that it is difficult to see how the heroic or altruistic personality type would become dominant in a species. Risking your own safety to help someone else survive means that you may not live as long, and therefore you will have fewer children. If you think about it, you might conclude that selfish rather than altruistic individuals are more likely to survive and have offspring, since selfish individuals expend all of their efforts fending for themselves and never risk their own survival trying to assist others. The evolutionary effects of this should be that, as altruistic individuals die out in their efforts to help their colleagues, they take the genes to be helpful and heroic with them. This would mean that we should expect that the proportion of individuals who take risks to come to the aid of others, or even those that expend otherwise needed resources to compassionately assist others, should steadily decrease. According to this reasoning, dogs, as a helpful, heroic, and altruistic species, would seem to be an evolutionary oddity.

Dogs do share their altruistic tendencies with humans. Altruism is relatively common in humans, and we only need an earthquake, flood, hurricane, or war to see people risking their own lives and spending their own resources to provide food, money, and medical supplies to help others survive. In virtually every major disaster we also see heroic acts in which rescue workers are killed or injured while assisting victims.

That is not to say that altruistic behaviors are *never* found in other species. In certain circumstances, other animals can act helpfully. For instance, there are confirmed reports of dolphins supporting sick or injured animals in their group by swimming under them for hours at a time and pushing them to the surface so they can breathe. Among baboons, when a predator, like a leopard, threatens the pack, dominant male baboons will hang back, defending the rear as the troop retreats. In a similar manner, when predators approach a herd of zebras, the adults will move back to the rear of the herd to place themselves between the threat and the younger, more vulnerable animals in the group. Yet such behaviors are sufficiently rare that they are quite striking when they do occur.

When we look at species more closely related to dogs, we also find that altruism is uncommon, but it does sometimes appear. In wolves, their predatory nature often overwhelms any altruism. A wolf that shows weakness or injury may become the target of attack by other wolves. The one situation in which wolves and wild dogs regularly show altruism is in sharing resources. Their sharing is not indiscriminate, however. Obviously, adults will share food with puppies and adolescents who are too young to hunt. Yet in some circumstances, hunters bring meat back to members of the pack that were not taking part in the kill. Usually these other pack members were not present because they were guarding young cubs near the den site. A skeptic might dismiss this as altruism and view the shared food as simply "wages" for the "babysitters" left at home while others hunted.

An interesting example of domestic dogs sharing resources when there appears to be no hint of any service being performed in return comes from Lynchburg, Virginia. The report dealt with a foxhound known as Old Red. He and a collie named Pete shared the home of A. A. Babcock. One day Pete disappeared and did not return for dinner. Babcock searched a while for the dog, but simply assumed that he had wandered off and would return as he had done once before. Old Red, however, started acting strangely after Pete's disappearance. One of the odd things that he started to do was to take chunks of his food and disappear from sight instead of eating them immediately as was his usual habit. Babcock became curious about this behavior and decided to follow the hound to see what he was doing. Red wended his way about a quarter of a mile (one-half kilometer) away from home.

There Babcock stumbled across an old mine pit, and trapped in the hole, suffering from what later proved to be a broken leg, was Pete. Around the collie was also evidence of a number of meals that had been brought to him over several days by his companion Red.

Altruism and Family's Genes

Darwin himself recognized that a single individual risking his own safety to help another person goes against survival of the fittest and places the altruistic person at a definite disadvantage. In his 1871 book *The Descent of Man,* Darwin noted that "he who was ready to sacrifice his life, as many a savage has been, rather than betray his comrades, would often leave no offspring to inherit his noble nature." But when we consider the survival of the group, rather than that of the person, altruistic behaviors become sensible. Darwin thus continued his discussion by noting that the presence of individuals who are altruistic and self-sacrificing might improve the likelihood that the individual's family, group, tribe, or species would survive. He suggested that "a tribe including many members who were always ready to give aid to each other and sacrifice themselves for the common good, would be victorious over most other tribes; and this would be natural selection." This selection would be for a genetic makeup that includes the gene or genes that make individuals more likely to be altruistic.

There are some complications, however. Suppose that we have a gene that causes its owner to behave altruistically toward other individuals by acting heroically or by sharing food with them. Animals without that gene are selfish, keep all their food for themselves, and don't take risks to help others. In addition, the selfish ones will sometimes get handouts from the altruists or be saved by the heroic actions of someone who has that gene. This looks clearly like the selfish individuals have a clear advantage and they should survive to have more offspring. This should then cause the species to evolve into self-interested and self-centered animals. However, there is a way for altruism to survive and flourish.

Suppose that altruists are discriminating about whom they share food with and whom they wish to risk their lives for. Suppose they do not help just anybody, but are more likely to be altruistic when their own relatives are involved. This obviously changes things since relatives are genetically similar, meaning that they share certain genes with one another. Thus, when an individual who is carrying the altruistic gene saves the life of a relative, there is a reasonable likelihood that the individual who is saved is also carrying copies of that gene. How likely this is depends on how closely the individuals are related. This means that genes for altruism can spread by natural selection. Although the gene causes an individual to behave in a way that reduces the individual's own fitness, it will increase the fitness of his relatives, who will have a greater than average chance of carrying the gene themselves. In the long run, the overall effect of heroic behavior may be to increase the number of copies of the altruistic gene found in the next generation, which in turn will increase the probability of helpful behaviors in the group or species.

As an example, suppose that you are a blackbird who suddenly notices the approach of a hawk. If you sound the alarm by giving a warning call to your neighbors, you place yourself at risk since that call will also make the hawk aware of your location and thus make it more likely that you will be the one who is attacked. However, even if you are killed because of your heroic raising of the alarm, you might save a reasonable number of close relatives in the vicinity. This means that the gene or set of genes responsible for triggering that alarm call behavior would still be successful and survive. Notice that this is a "gene's-eye view of evolution." It suggests that evolution is the result of competition among genes, each of which is trying to achieve an increased representation in the total gene pool of a species. In this argument, individuals are simply "vehicles" that genes have built to aid them in reproducing. Although altruism might be bad from the point of view of a single person, from the gene's point of view it makes good sense. If the goal of the gene is to end up with the maximum number of copies of itself in future generations, one way to do that is to behave altruistically toward other individuals who also carry the gene.

This theory would predict that the amount of help that you offer to others would be greater for those with the closest relationships to you since they would have the most genes that are similar to yours. Research generally confirms this. The most obvious example is the risk and suffering that mothers are willing to endure to defend their offspring and keep them safe and healthy. This is the case for virtually all warm-blooded species. Another example comes from certain bird species, where there are some birds that give up their breeding opportunities in order to help provide food and defend chicks that belong to other birds. Research shows that these "helper" birds are much more likely to help close relatives raise their young than they are to help unrelated breeding pairs. Additional support for this conclusion comes from studies of Japanese macaque monkeys. These monkeys have been known to act heroically by defending others from attack; however, monkeys are most likely to put themselves at risk if the threatened individuals are close kin.

Can People Be Kin to Dogs?

If altruism were based on kinship, this "survival of the genes" explanation of why dogs help people would seem to fail. Obviously, a dog and a human cannot be part of the same family in the sense of sharing genes because dogs and people are not even the same species. However, even when we are considering helpful behaviors between animals of the same species, there is no requirement for them to have the ability to discriminate relatives from non-relatives. While mothers can certainly know their exact relationship to their children, everyone else seems to use certain rules of thumb. For example, individuals who are raised together are more likely to be kin than non-kin. The reverse of the process works as well, so that living in the same household with another individual is often enough to cause those individuals to treat each other more like family. It is likely that dogs are affected by the same mechanism—simply living with a particular group of humans causes them to view those people as pack mates and perhaps even family members.

The species difference between humans and dogs should not be a major problem for the theory. If dogs are properly socialized as puppies, with much handling and interaction with humans, we know that they tend to accept people as pack members. It is also true that most members of a wolf pack are genetically related. Dogs communicate with people in the same manner as they communicate with other dogs, develop dominance relationships with people, and so forth. Now combine these facts. Since dogs seem to accept us as pack members (which are likely to be relatives) and they have grown up with us and live in the same social unit and location as us (just like relatives), it would seem sensible for dogs to treat us as if we were relatives and act altruistically toward humans.

Some dogs seem to define family and pack members based upon distinguishing features. It is not unusual to find dogs that seem to develop a sense of kinship or fondness for certain people based upon their age, race, sex, or some other distinguishing characteristic. Take the case of Laska, a handsome white Samoyed who lived the first two and a half years of her life as the sole companion of an elderly man. When her owner suddenly died, Laska became grief-stricken. She moped, would not interact with other humans, and barely ate enough to sustain her. Ultimately, she was adopted by the Richardson family in Brighouse, Yorkshire. A lot of care and attention, plus the companionship of the family's other pet Samoyed, Emma, brought her out of her depression and for a year she lived a happy and contented life. Then one evening Laska disappeared; apparently she left through the garden gate that one of the Richardson children had left open. Despite the fact that it was one of those cold and wet nights that is typical of the end of September in Yorkshire, a search was organized. However, after five hours the inclement weather defeated them and the family sadly went to bed after leaving a report with the police.

Meanwhile, some 8 miles (13 kilometers) away, in Bradford, the police were engaged in another search. Norman Stephenson, a frail eighty-one-year-old man who was suffering from increasing memory losses, perhaps due to the beginning stages of Alzheimer's disease, was reported missing by his wife Mary. She was extremely worried because the hour was late and the weather was bad, and he had left their home without his coat. The night patrols did not find the old man.

The next morning the search for Stephenson started again. It was the barking of a dog that attracted searchers to the edge of an embankment some 4 1/2 miles (7 kilometers) from his home. At the foot of the slope the man they were looking for lay unconscious. However, snuggled up to protect him was Laska, who had huddled near him to share her body heat through this sixteen-hour ordeal. Suffering from exposure, Stephenson was rushed to the hospital, where he gradually began to recover. The consensus was that the old man would have certainly died if the dog had not stayed and provided critical body warmth and protection.

At one level, Laska's behavior was understandable. Her Russian ancestors who were used to pull sleds were known to have had a protective instinct toward their owners. There are a number of reports of how Samoyeds have helped to keep their masters alive in the extreme cold of the Arctic, by wrapping themselves around their human family members to provide warmth. Yet Norman Stephenson was a stranger, not Laska's master. In this case, the dog's protective actions were most likely triggered by her deep affection for her original master and the fact that this human was male and elderly as well, and must have evoked a sense of recognition.

Selfish Altruism

A cynic might say that applying the label of altruism to canine behavior is wrong, and the only reason that dogs help is because it benefits them to do so. In essence, the idea is that the dog effectively thinks, If I help you out in the crisis, then when I am in trouble sometime in the future, you'll help me. The cost, in terms of reduced evolutionary fitness because of the risks involved when the animal behaves heroically could thus be offset by the likelihood of this return benefit. If the benefits outweighed the risks, this could permit altruistic behavior to evolve by natural selection. We could call this *reciprocal altruism*, or just a version of "You scratch my back and I'll scratch yours." In effect, this suggests that altruism and heroism are just forms of delayed self-interest.

This reciprocal altruism has some advantages as an explanation for why dogs help since there is no need for the dog and the person to be relatives, nor even to be members of the same species. The only requirements are that individuals should interact with each more than once, and have the ability to recognize those individuals that they have interacted with in the past. The reason for these requirements is that if individuals interact only once in their lifetimes and never meet again, then obviously there is no possibility of return benefit, which would mean that there is nothing to be gained by behaving altruistically. This also suggests that individuals who don't act helpfully toward others can be recognized and punished by not being helped at a later time when they are in need.

There is some evidence that reciprocal altruism does occur in nonhuman animals. Gerald Wilkinson of the biology department of the University of Maryland demonstrated how it works in vampire bats. It is quite common for a vampire bat to fail in its attempt to feed on any given night. While an individual bat can survive for a day or two without feeding, it can't last much longer than that. When bats do successfully feed, they tend to gorge themselves beyond their own immediate needs. The altruistic behavior occurs when, after returning to their roost, a bat may donate blood (by regurgitating it) to less successful members of their group who have failed to feed that night. This serves to save the other group members from starvation. Wilkinson's research demonstrated that a bat will tend to share its food with its close associates, which is what we would expect if familiar individuals were seen as family and the attempt was simply to preserve common genes. However, there was also evidence of reciprocal altruism since bats were also more likely to share with other bats that had recently shared with them.

It is certainly the case that helpful, useful, and heroic dogs get reciprocal benefits as a result of helping people. They get good housing, food, protection, and medical care, and their puppies are also cared for (which helps to preserve the genes of the helpful dog). However, it seems unlikely that prior to a heroic act Rover is consciously working out the costs and gains, with reasoning like, "If I help this human survive, he may pay me back by helping me to survive longer, and my puppies will get to propagate and preserve my genes."

Altruistic Dogs Create Altruistic People

It is likely that dogs heroically help humans because of a combination of factors. It works out something like this: Dogs view humans as part of their family, and that sense of kinship increases the likelihood that dogs will help humans. Dogs are also more likely to help those humans that have helped, or been kind and supportive to them in the past. If the human they help survives because of the dog's heroic actions, then it becomes more likely that that dog and its offspring will be cared for. In this way, the dog's altruistic genes will also survive and multiply in the future. Thus, the dog's altruism affects the genetic makeup of future canine generations.

An interesting and unexpected consequence of canine altruism and heroism is the possibility that it has had an effect on human genetics and evolution. Since it is more likely that dogs will assist those people who have been kind to them in the past, that would increase the likelihood that altruistic people who helped dogs in the first place will survive longer. This means that the dog's altruism toward people makes it more likely that altruistic genes in humans will also be passed on and multiply in the future. This is a much more extreme version of taking a gene's-eye-view of the process, since it says that the gene really doesn't make much of a distinction as to species as long as it, and the characteristics that it helps to create, continue to survive.

We can summarize the situation in everyday language that does not depend upon discussions of evolution or genes: dogs help us because they view us as family and because they believe we would help them if the situation were reversed.

Has altruism really helped the survival of dogs? From our previous discussion of attempts to tame wolves, we know that wolves remain fearful and aggressive toward humans despite our best efforts to build a bond with them. It is thus obvious that while wolves may occasionally act helpfully toward other wolves, they don't act helpfully toward humans. Remember that the evolutionary success of a species can be measured by the number of individuals who survive and reproduce. If that is the case, then it is important to note that there are a thousand times more dogs in the world (about 400 million) than there are wolves in the world (about 400,000). To the extent that their heroic and helpful behaviors have contributed to our keeping dogs, this would suggest that it is a very successful evolutionary mechanism.

Although the creation of dog breeds has involved selectively breeding animals that have specific characteristics that we desire, there is no evidence that anyone has tried to specifically breed dogs that are heroic or altruistic. However, that very thing may have been going on in a subtle, unconscious way. A dog that has saved your life or that of a family member will be cherished, protected, and might be more likely to be bred several times so that others can receive puppies similar to your wonderful dog. In this way the altruistic genes get passed on and ultimately might work their way through a particular breed of dog. This raises the question of whether there are specific breeds of dogs that are more likely to be heroic, and more likely to try to help human beings.

Are There Heroic Breeds?

One can, of course, speculate in advance of the data about which breeds might be most likely to act heroically in aiding a human. Dog breeds that work closely with people and have to respond to their commands and signals will have the highest probability of viewing their human masters as family members, and thus be expected to help in a crisis. Obviously, this would include herding dogs, many of the sporting dogs, and some of the working dogs. Terriers and hounds are expected to act more independently, and might not form the close association needed to trigger altruistic behaviors. Size may also be a factor, since a 5-pound (2-kilogram) Yorkshire terrier will simply not have the mass to drag you out of a lake if you are drowning, knock you out of the path of an oncoming vehicle, or chase away a bear that is threatening you. A small dog might have the heart, but simply not the mass to do what is needed.

An example of this is the sad story of a four-year-old Chihuahua, owned by the Furman family who lived in Enfield, New York. The dog lived up to its name, Tiny, by being small even for one of her breed. Nonetheless, she was not lacking in courage and was known to sound the alarm when people approached, and to stand in front of strangers barking her defiance while trying to protect her home. One day when the family was out, burglars broke in. Tiny rallied to the defense, but her size made the effort futile. The burglars were annoyed by her attempts to stop them and the noise she was making so they snatched her up and threw her into a clothes dryer. They then turned it on and left her there to die a horrific death. A postscript to this grim story is that apparently Tiny's defense of the house was sufficiently noisy and vigorous so that the burglars must have feared discovery and they fled before they had an opportunity to do more than ransack the laundry room, which was the way they had broken into the home.

To determine scientifically which breeds are most likely to help, we can return to the 1,006 cases of dog heroism that I collected from news reports. One criterion for selecting particular cases was that something had to be said about the dog's breed. Of the cases that I collected, 657 were purebred (or at least of a definable breed group as far as the reporters were concerned) and 349 were listed as mixed breed. This means that approximately one-third of the heroic actions of dogs were attributable to mixed-breed dogs. It is hard to classify or to analyze the results of mixed-breed dogs (do we classify a Labrador retriever and collie cross as a sporting or herding breed, or a little of both, or . . .), so I have chosen to do the analysis on the 657 that were identifiable at least as to breed group. This means that a dog identified only as a "terrier" or a "spaniel" can be counted for the present purposes, although it would have been better if we knew which kind of terrier or spaniel was involved. Nonetheless, this provides us with enough information to classify these heroic dogs into the seven breed groups used by the American Kennel Club, namely sporting dogs, hounds, working dogs, terriers, non-sporting dogs, toys, and herding dogs.

Looking at the total number of heroic dogs, the group that ranks the first is the herding group which accounts for 36.5 percent of the reported rescues. For the purpose of this analysis, we can consider this group as including several types of dogs. Two groups are quite distinct; first are the collie-type dogs (for example collies, border collies, Shetland sheepdogs, and bearded collies), second are the herd guarding dogs (such as German shepherd dog, Belgian sheepdog, Malinois, and Tervuren), which then leaves a third more heterogeneous group (such as the corgis, Australian cattle dog, and Bouvier des Flandres). Remarkably, over the years surveyed, German shepherd dogs accounted for 16 percent of all rescues reported, followed by the collie-type dogs with 13.3 percent. There is an interesting sidebar to the data on German shepherds in that they seem to have changed over time. If we look at the percentage of heroic rescues that they were involved in over the period between 1938 and 1984, they were responsible for 18.1 percent of all rescues while in the period 1985 to 2005, they account for 10.6 percent of all heroic acts. The fact that in recent years German shepherds seem less likely to help people, despite the fact that their popularity has not diminished much, suggests that there has been a significant change in the breed, perhaps in their temperament, since some reports seem to suggest that today's German shepherds are a bit more skittish and fearful than in former years. It also may be due to changes in their physical abilities since, at least in North America, the fashion of breeding for strongly sloped backs seems to have greatly reduced their speed, power, and agility by restricting free movement of their rear ends. Whatever the reason the accounts of heroic actions by this breed are now fewer than they used to be. Even given that change, however, the German shepherds still are way ahead of many other breeds in the number of rescues that they have made in recent years.

The group of breeds that rank second is the sporting group with 27.2 percent of the total rescues. This group includes the retrievers (for example Labradors, goldens, flat-coats, and Chesapeakes), the setters (such as Gordon, Irish, and English), the spaniels (including cockers, field, Sussex, and Brittany) as well as the pointers. The retrievers account for the largest number of rescues in this group (with 6 percent), spaniels next (with 3.7 percent), and then setters (3 percent).

The working breeds rank third in total rescues with 17.8 percent of total rescues. This is a somewhat mixed group with dogs that are primarily guard dogs (including Doberman pinschers, bullmastiffs, Great Danes, and Rottweilers), so-called husky-type dogs that were used as draft animals (including malamutes, Samoyeds, and Siberian huskies), and some mixed-use dogs (most notably the Saint Bernard, Newfoundland, and the schnauzers). In this group Newfoundlands are most often involved in rescues (5.6 percent of the total) with Saint Bernards not far behind (4.3 percent). There is also a significant appearance made by huskies (2.3 percent) and Great Danes (2.1 percent).

These three groups—herding, sporting, and working—are the ones that we originally predicted would be the most likely to be involved in rescues, since they are the breeds that actually work in close association with people. Such close associations require the dogs to pay close attention to humans, which means that they are more likely to spot trouble if it occurs. However, this association also provides the conditions that are most likely to produce feelings of family-like kinship. Together these three groups account for 81.5 percent of the total number of rescues.

One surprise was that the terrier group accounted for nearly as many heroic acts as did the working group with 16.4 percent of the total. Perhaps it is the vigilance, excitability, and persistence of terriers that make them such good watchdogs, ready to warn families about fires, gas leaks, or people in trouble. Take the case of Lilly, a bull terrier owned by Jimmy Farrell. After working a double shift, Jimmy had toppled into bed to get some much-needed sleep. His mother, Joan, had gone downstairs to put a load of laundry in the wash, and as usual, Lilly had tagged along. As she entered the laundry room, Joan had a heart attack and collapsed. Lilly was startled by Joan falling down, and recognizing something was clearly wrong she went to Jimmy's bedroom and began barking. The exhausted Jimmy was reluctant to respond, since he believed that Lilly only wanted to come to bed with him as she usually did. Lilly was persistent, and when the door didn't open she barked more frantically and began banging her head against his closed door repeatedly. Her frantic behavior finally forced Jimmy out of bed. He opened the door, and crawled back under the covers expecting Lilly to follow. Instead, she pulled at the blankets and tugged on his arm until he finally sensed something was wrong. When he got out of bed Lilly led him to the laundry room, barking at him whenever he paused, and there Jimmy found his mother unconscious on the floor. He immediately called for help and paramedics arrived in time to stabilize Joan and save her life.

The terrier group can be roughly divided into the working, vermin hunting, terriers (such as Airedales, cairns, West Highland white, soft-coated wheatens, and Scottish) and those that are classified in popular lore as being bull terriers based upon their appearance (American Staffordshire, Staffordshire, and of course bull terriers). It is the fox terrier (smooth and wire-haired) that seems to be the most heroic of the terrier group, accounting for 3.8 percent of all heroic acts. Bull-type terriers (like Lilly) make a respectable showing with 2.6 percent of heroic acts.

After the terrier group there is a marked decrease in the number of rescues per group. The nonsporting group accounts for only 8.9 percent of total rescues. This is a remarkably mixed group that contains such diverse breeds as the Dalmatian, chow chow, Chinese shar-pei, Lhasa apso, Boston terrier, and schipperke. Of this group, the breed that accounts for two-thirds of the rescues made by non-sporting dogs is the bulldog (3.3 percent of the total). The only other dog that makes a significant appearance in this group is the poodle, accounting for 1.2 percent of the heroic acts.

The hound group includes the sight hounds (such as the greyhound, whippet, Afghan, Irish wolfhound), the scent hounds (like bassets, beagles, bloodhounds), and a mixed group (including foxhounds, coonhounds, Rhodesian ridgebacks, and dachshunds). Together this group accounts for only 3.8 percent of total rescues.

The group that we expected might be underrepresented in heroic acts simply by virtue of the diminutive size of the dogs is the toy group. It includes dogs like the Chihuahua, Maltese, papillon, Pekingese, Pomeranian, and Yorkshire terrier. Although toy dogs are extremely popular, they account for only 2.4 percent of total rescues.

A Breed for Every Crisis

It is interesting to see which kinds of rescues the various breeds are involved in. When it comes to sounding the alarm to alert humans to a problem, herding dogs and terriers seem to be the most alert and insistent. Herding dogs have a few more rescues where fire and smoke presented the danger to which their owners had to be alerted, while terriers had the edge when the problem involved gas leaks and exposure to toxic chemicals. Both groups used much the same tactics in sounding the alarm, namely: barking, jumping, pawing, and frantic racing back and forth between the source of the problem and their human family members. As might be expected from the overall results, the herding dogs most likely to sound the alarm are collies and German shepherds, and the terrier most likely to detect a problem is the fox terrier.

When it comes to protecting people physically from some sort of attack, again we find a large number of herding dogs, surprisingly followed by terriers and only then the working dogs. In the case where an individual is being attacked or threatened by a human being, German shepherds seem to automatically spring to the defense, but all terriers (particularly fox and bull-type terriers) seem to try their best to protect their loved ones from menacing people. Another surprise was that although the smaller terriers often suffer grave physical injury during such confrontations, they seem to be remarkably successful in driving off attackers. As might be expected the various guarding breeds of working dogs (such as Dobermans, Rottweilers, and Great Danes) that are often found interposing themselves when humans are the source of a threat. When the threat is an animal, however, the situation is different. Here sporting dogs are most likely to intervene, particularly the retrievers. This might well be simply because it is the sporting dogs that are most likely to be out in the wilderness areas where wild animals are to be found; however, there are numerous reports of sporting dogs intervening to stop another dog from attacking a human family member or to defend a person from an angry domestic bull or horse. The non-sporting group also makes a significant contribution in cases where a dog physically intervenes to protect people from attacks due mostly to the lumbering intervention of courageous bulldogs.

When it comes to summoning help for people that are injured, ill, trapped, or lost, herding dogs again are involved in the most rescues. The sporting dogs are the next most frequently reported as engaged in this kind of helping behavior, while quite close behind in frequency are the working dogs. However, the particular circumstances of the problem determine which dogs are most likely to bring help. Newfoundlands, St. Bernards, and huskies are the working dogs most likely to bring help to people who are lost or trapped, although the collies and German shepherds are prominent in this kind of rescue.

When it comes to bringing aid to people who are injured or sick, the sporting dogs make a strong showing with retrievers most likely of the sporting dogs to do this, followed by spaniels. A typical example of this kind of rescue comes from Bullet, an aging golden retriever who alerted his mistress, Pamela Sica, when her baby began gasping for breath. She described it this way:

"I was in the kitchen making the bottle. He [Bullet] *was in the bed-*

room with my son. My husband went into the shower. Bullet was still lying down and I guess when the baby was making the sounds, he came running down the hallway into the kitchen. And he kept barking, and I was still making the bottle and I asked him if he wanted to go out, and he kept barking and turning around and going into the hallway. Then I finally went into the bedroom, and that's where I found my son. And he had his head back, and he was gasping for air. With that, he was turning a shade of red too, like, purple to blue.

"I screamed for my husband. He came out of the shower and with that, he turned the baby upside down. He thought that I fed him, so he thought he was choking. So he hit him a couple of times on the back and it didn't do anything, and he turned him around and started to rub his chest and do CPR."

Pamela immediately called for help and the emergency medical team arrived quickly. They stabilized the child's breathing. The baby, Troy Sica, was found to have had pneumonia in both lungs. It was only Bullet's early warning and the father's artificial respiration measures that kept the baby from dying that night.

This story is typical of the way that retrievers and other sporting dogs summon help for people who are injured or ill. Such physical crises also are the one circumstance where hounds seem to show their heroism.

Some of the most spectacular reports of dogs physically intervening to rescue humans are the water rescues—people at risk of drowning because they had fallen into the water or gone through broken ice. Such people are usually rescued as a result of the dog supporting them so that they could stay above water and breathe, or when the dog physically drags them to shore or shallow water. While the herding dogs again are frequently mentioned in such rescues, the working and sporting dogs are also often involved in such reports. Among the working dogs, the Newfoundland and the Saint Bernard together account for close to one out of every five such rescues.

Newfoundland dogs are famous for water rescues. One could write an entire book about their exploits in this regard. The fame of Newfoundlands as water rescue dogs was virtually insured by a dog named Milo, owned by George B. Taylor, the keeper of the Egg Rock lighthouse off shore from Nahant, Massachusetts. Milo was a black and white Newfoundland, and became famous when several newspapers ran articles describing how he had saved the lives of several children caught in the unpredictably strong currents of the surrounding water. The artist, Sir Edwin Henry Landseer, was attracted by these stories, and painted a pair of portraits of Milo. One was quite traditional with Milo resting on the dock and it bore the title *A Distinguished Member of the Humane Society,* and the other showed the great dog, looking exhausted and concerned, having just pulled Taylor's young son, Fred, from the icy water. This painting, titled *Saved,* became famous across America, and resulted in black and white Newfoundlands ever since being called "Landseer Newfoundlands" after the artist who portrayed them so dramatically.

An interesting sidelight is that this heroic and altruistic behavior of rescuing drowning people has now been turned into a sport. Although the basic behaviors seem to be instinctive, they can be honed so that the dog can perform many forms of water rescue. Tests have been devised and dogs that pass them can earn titles such as "Water Dog" or "Water Rescue Dog." The dogs are required, among other things, to "rescue" an "unconscious" person and bring them to shore. They are also required to determine which of several people in the water is in trouble (apparently floundering around) and to rescue that person, ignoring others that appear not to be in distress. Later on, they will tow lines out to "stranded boats" and help drag them to shore as well.

Returning to the data on rescuing drowning people, the sporting group is also mentioned frequently in that regard. It is probably to be expected that the water-loving sporting dogs, namely the retrievers, spaniels, and setters (in that order) are responsible for the next largest proportion of people rescued from drowning. One surprise for me was that there were actually three cases where bulldogs were involved in saving someone from drowning. Given their dislike of water and poor swimming ability, I went back and checked the facts a bit closer. It turned out that two of these involved people who went into the water because they crashed through thin ice, so the rescue involved pulling rather than swimming. That still leaves one case where a bulldog seems to have had to swim, at least a short distance, to save a child, which I still find astonishing.

Animals Rescuing Animals

There is one final category of rescue that I did not include in the rescue statistics, but is of interest. This is the category where dogs rescue other animals. If I had included it in the analysis of dog heroics, it would have accounted for close to 5 percent of the reported incidents. Some aspects of dogs rescuing other animals are consistent with the idea of helping preserve the genes of related individuals, since more than half of the rescues (56 percent) involve dogs rescuing other dogs. However, 22 percent of the rescues involved dogs rescuing cats, which certainly don't share canine genes. Among the many other species saved by the heroic action of dogs were horses, birds, cows, a rabbit, a hedgehog, and even a family's pet bear. Most rescues of the larger animals involved alerting humans to fires or trapped animals, although in two instances dogs actually rushed into burning barns and chased horses out to safety.

One can't rule out maternal instinct in such cases, since one-third of the dogs rescued by dogs were puppies, and these were saved by female dogs (although not their mother). In addition, approximately half of all of the cats saved were kittens; again all of the rescuers were female dogs.

The distribution of breeds involved in these altruistic heroics is somewhat similar to what we saw in the rescue of humans, although the differences among breeds are less extreme. Herding dogs (almost all German shepherds or collies) account for 18 percent of the rescues of animals. Close behind are the sporting dogs, with 13 percent of the animal rescues, and here retrievers are responsible for two-thirds of the heroic acts in the sporting group. Terriers also account for 13 percent of helping actions (mostly alerting to a problem by making a lot of noise). Although toy breeds are not found at all in the reports, it was a surprise to find that close to 6 percent of rescues of other animals were made by hounds.

Not all rescues of other animals seem to be motivated by altruism, however. Consider the somewhat bizarre case of Fray, a golden retriever owned by William Brown of Norfolk, Virginia. In addition to Fray, Brown owned two other dogs, a French bulldog named Louis and a pug named Pinzo. Brown's home was on a rocky beach, and a pier extended out into the water where he would sometimes sit on while he was fishing. The three dogs would usually sit or lie on the pier not far from him. One day, as Brown sat lazily with his fishing line in the water he heard a splash. He looked around and saw Fray looking over the edge of the pier into the water below. The tide was coming in, and the water was rough, but Brown saw Pinzo's head bobbing up and down as the pug valiantly tried to stay afloat. Before he could think of anything to do, the golden retriever leapt into the water and grabbed Pinzo like a retrieving toy, pulling him to the shore. Brown rushed to the dogs and found that the pug was sputtering but otherwise uninjured. Brown was ecstatic at Fray's heroism and made a huge fuss over him. The dog was given a big piece of meat as a reward as well.

A few days later the scene virtually repeated itself, only this time it was Louis, the bulldog, who ended up in the water and had to be rescued by Fray. The retriever again was the focus of great attention and once more was rewarded with a special food treat. Two days later Pinzo the Pug was again plucked from the water by Fray, and only a day after that it was Louis who again had to be saved by the retriever. At this point Brown was becoming concerned. Neither of his two small dogs was a strong swimmer, and both seemed to be becoming accident prone when out on the pier. If he had not had Fray to rescue them, he felt sure that one or both of them would have been lost.

The next time he went out on the pier to fish with the dogs, he made sure that he glanced frequently in their direction. He wanted to find out what was attracting the small dogs to the edge of the pier and why they were falling over its side. After a half hour had passed, Brown noticed that Fray had stood up. Brown quickly looked and was relieved to see that the other two dogs were sleeping peacefully. Now Fray quietly walked over to the pug, grabbed his collar in his teeth and with a snap of his head sent the little dog over the edge and into the water below. Fray looked over the side at the struggling dog, gave a single bark, and leapt over the side to once again "rescue" his housemate. It is difficult to know what Fray's motivation was. Perhaps it was simply because he had learned that by "saving" these little dogs, he got attention, affection, and special treats. Perhaps he was merely bored at sitting and watching his master fishing, and this was a means of playing retrieving games, only here with live toys to fetch from the water, rather than balls or bumpers. However, it is certain that this was neither altruistic nor heroic behavior on Fray's part since he was creating the problems which called for his intervention. Whatever urges turned this dog into a counterfeit hero they were clearly based upon self-interest rather than a desire to help others.

This whole scenario had the elements of an intelligence test, though. Fray was smart enough to devise this new game, while the other dogs were either willing to go along, or not clever enough to figure out how to avoid being part of it.

AFTERWORD

While writing this book I was occasionally reminded of a conversation that I had with Donald O. Hebb in the early 1980s. You might remember that he was the psychologist whose research first gave us an understanding about how interactions with the environment actually help to change the structure of an individual's brain, and this eventually led to the environmental enrichment procedures that we use to create superdogs. He had come to give a series of lectures at the university but was now finished. We were sitting in a colleague's living room chatting. As I reconstruct his comments in my mind they started after I mentioned to Hebb that I was thinking about writing a book on the intelligence of dogs and then perhaps following it up with a book on the personality of dogs.

He smiled and said in his gentle Nova Scotia accent, "*You'll get into trouble with your scientific associates if you use words like 'personality' or 'intelligence' to describe the results of research on dogs. They'll accuse you of 'anthropomorphizing.' Most likely they will assume that you are some softheaded thinker who believes that animals are pretty much just fur-covered humans that think and act the way people do. It'll probably do your career more harm than good.*"

He took off his rather severe-looking glasses and wiped them absentmindedly as he continued:

"Back in the 1940s I worked for something like two years in the Yerkes Primate Research Laboratory trying to describe the temperaments of some of the captive chimpanzees that they used for behavioral research. At the time there was official prohibition against using anthropomorphic descriptions in the scientific reports of any research completed there. I certainly never would have dared to use the word 'personality' in talking about a chimp. I was told that even to say something like 'that animal was afraid' was not good practice since it hinted that the animal felt fear the way that humans do. Instead I was expected to simply describe the conditions that might have stimulated behaviors and then to describe those behaviors objectively. Like when I showed chimps a life-sized model of a human head with no body attached, I was supposed to say that the animal ran to the back of the cage and cowered down and screamed or whimpered, instead of simply saying that 'the animals were frightened by sight of certain unusual or strange objects.' To hint that the animal was 'afraid' would have been considered to be anthropomorphizing.

"Well, the truth of the matter is that when I did try to objectively describe the temperaments and behavior patterns of the animals without using the words we use to describe human emotions, all that I ended up with was a giant mess. I mean, all that I had created was this immense list of specific acts and specific situations. You really couldn't find any order, pattern or meaning in that kind of data. Unfortunately, at the practical level, focusing attention only on specific acts and behaviors was also a bit dangerous. A couple of times I was so caught up in recording behavior descriptions that I missed the animal signaling that it was annoyed or unhappy with me, and I nearly had some fingers bitten off—or worse.

"While I was going through all of this, I couldn't help but notice that the staff or keepers—you know, the people who cared for the animals on a daily basis and who don't have advanced degrees and don't need to worry about research purity—didn't seem to have any problems. They used the same kind of intuition that we normally use when we observe the behavior of people. Because of that they could describe one animal as having a 'dominant personality,' another as being 'nervous,' another was considered to be 'a friendly beast,' still another was 'shy' and there was even one that they claimed was 'bashful.' These were clearly anthropomorphic statements which suggested that, like people, the animals had

distinct and individual personalities and that you could use their personalities to predict the animals' future behaviors.

"If I were trying to be a totally objective researcher, especially given the scientific attitudes of that time, I should have rejected their statements as speculative, anthropomorphic, nonsense, but to be truthful I didn't. You see, the words that the animal care staff used to characterize the behaviors of those animals were useful and helpful. When they described an animal in this way to a newcomer, or even a psychologist who was not too arrogant to listen, that 'personality' information allowed that person to predict how the animal would respond and to safely interact with it.

"Their anthropomorphic descriptions obviously suggested that each animal had certain attitudes and behavior predispositions. It also implied that each animal experienced predictable emotional changes. Whether this is so or not, I can't say, but it did provide an intelligible and practical guide to the behavior of those animals. It clearly worked with the chimps, and I think that it should work with any animal including dogs."

He put his glasses back on and looked directly into my eyes. *"I'd rather like to see those books after you've written them. I know that dogs have intelligence, and I'm sure that I can be convinced that they have something very much like personality, at least in the sense that they have predictable tendencies to act in specific ways which are unique to that individual. I would like to see if you can put together the evidence to scientifically establish those points."*

As often happens, the data collection took longer than I expected, and it was many years later that these books were finally realized. Unfortunately, Hebb, who was around seventy-five years of age at the time of our conversation, died before he had the chance to see the books and offer his comments on the final product. However, his notion that a bit of anthropomorphizing, or thinking about animals in the same way we think about humans, has turned out to be useful. Not that I am suggesting that dogs are little four-footed people in fur coats, but rather that thinking about the behaviors of dogs in the same way that we think about the behaviors of humans can help us to understand and predict the behavior of dogs.

People have personalities, and these show up as consistent predispositions to act in certain ways. We can determine some aspects of the personality of a person by observing their behavior, knowing their life history, or looking at the results of scientific personality tests that an individual has taken. An estimate of an individual's personality can serve as a practical starting point for predicting how they will react and behave in a variety of situations.

If we accept the fact that dogs have personalities as well, in the sense that they also have consistent predispositions to act in certain ways, then we can use the same kind of thinking that we use with people to predict canine behaviors. To understand a dog we can observe his behaviors, look at scientific information that is available based on the dog's breed (or dominant breed in a mixed bred), learn about that dog's history and rearing, and look at the results of a personality test (like the Dog Behavior Inventory in this book). Together these can give us a reasonable basis to predict how a dog will act, since the personality traits that dogs have are similar to those that humans have. It will also explain why individual dogs may respond in different ways to the same situation, since, like humans, each has its own particular personality.

At the risk of anthropomorphizing one more time, I would like to suggest that what American author and advertising executive Bruce Barton said about people is true for dogs as well: *"If you have anything really valuable to contribute to the world, it will come through the expression of your own personality, that single spark of divinity that sets you off and makes you different from every other living creature."*

APPENDIX: PERSONALITY PROFILES OF INDIVIDUAL BREEDS

The table below presents the data based on ratings by ninety-six dog experts, on the personality traits of 133 individual dog breeds. I have also listed the group that each dog falls into so that you can refer back to the discussion in chapter 5. The five traits measured are:

Dominance/Territoriality: This is a measure of how assertive and possessive a dog is. Dogs high on this trait will make good watchdogs and guard dogs, but they also may be quite vigorous in guarding their own things, such as food and toys. This trait is also most closely associated with *aggression.*

Intelligence/Learning ability: This is a measure of how easily a dog learns and solves problems. It also indicates how easy it will be to teach the dog obedience commands and how to perform service activities. In humans, the aspect of personality associated with this trait is called *openness.*

Emotional reactivity: This is a measure of how quickly the dog's mood may change and how fluctuating his emotional state may be. A dog that is high on this dimension will be easy to excite but will also calm down easily. A dog that is low on this trait will be more difficult to excite but will also take much longer to calm down. In humans, the aspect of personality associated with this trait is called *neuroticism,* which is usually contrasted with its opposite pole, *stability.*

Sociability: This refers to how friendly a dog is, and how much a dog seeks out companionship. A dog high on this trait will happily greet any new person, while a dog low on this trait may appear shy and aloof and may prefer to be alone rather than with people most of the time. In human personality terms, this is usually labeled *agreeableness.*

Energy: This is a composite measure that looks at the dog's indoor and outdoor activity level. It also includes a measure of vigor, which is the amount of force and energy the dog will bring to common activities. Thus, a dog high on energy may tug strongly on the leash or quickly and forcefully grab a treat from its owner's hand, while a dog low on the energy dimension will be much gentler in such activities. In human personality systems, this is most closely related to the dimension *extroversion,* which is generally viewed as the opposite of *introversion.*

The Ranking System

The data have been divided in quartiles, which are 25 percent groupings. Thus:

Very Low is the lowest 25 percent compared to other dogs ranked for this trait (75 percent of all dogs score higher than this).

Moderately low is higher than the lowest 25 percent of the dogs but lower than 50 percent of the dogs ranked for this trait.

Moderately high is higher than 50 percent of the dogs but lower than the top 25 percent of dogs ranked for this trait.

Very high is the highest 25 percent compared to other dogs ranked for this trait (75 percent of all dogs score lower than this).

Please note that for the energy dimension the experts felt that all breeds were capable of periods of high energy, so that even the lowest rankings are not truly inactive dogs.

Breed	Group/ Function	Dominance/ Territoriality	Intelligence/ Learning Ability	Emotional Reactivity	Sociability	Energy
Affenpinscher	Companion	moderately low	moderately high	moderately high	moderately low	very high
Afghan hound	Sight Hound	moderately high	very low	moderately low	very low	moderately low
Airedale terrier	Vermin Hunter	very high	moderately high	moderately high	moderately high	moderately high
Akita	Spitz	very high	moderately low	very low	very low	very low
Alaskan malamute	Spitz	very high	very low	very high	very high	moderately high
American foxhound	Scent Hound	moderately low	moderately low	very high	very high	moderately high
American Staffordshire bull terrier	Fighting Dog	very high	moderately low	moderately low	very low	moderately high
American water spaniel	Spaniel	moderately high	moderately high	moderately high	moderately low	moderately high
Australian cattle dog	Drover	moderately high	moderately high	very high	very low	moderately high
Australian shepherd	Herding	very low	very high	moderately high	moderately high	very high
Australian terrrier	Vermin Hunter	moderately low	very high	moderately low	moderately low	moderately high
Basenji	Sight Hound	moderately high	moderately high	very low	very low	moderately high
Basset hound	Scent Hound	very low	very low	very low	very high	very low
Beagle	Scent Hound	very low	moderately low	moderately low	very high	moderately low
Bearded collie	Herding	very low	moderately high	moderately low	very high	moderately low
Bedlington terrier	Vermin Hunter	moderately low	moderately low	moderately low	very low	moderately high
Belgian Malinois	Herding	moderately low	very high	very high	very low	moderately high
Belgian shepherd	Herding	moderately low	very high	moderately high	very low	moderately high
Belgian Tervuren	Herding	moderately low	very high	very high	very low	moderately high
Bernese mountain dog	Draft	moderately low	very high	very low	moderately low	very low
Bichon frise	Companion	very low	very high	moderately low	very high	moderately high
Black and Tan coonhound	Scent Hound	moderately high	very low	moderately low	moderately high	very low
Bloodhound	Scent Hound	very low	very low	very low	very high	very low
Border collie	Herding	moderately low	very high	very high	very high	very high
Border terrier	Vermin Hunter	moderately low	very high	very low	very high	moderately high

Breed	Group/ Function	Dominance/ Territoriality	Intelligence/ Learning Ability	Emotional Reactivity	Sociability	Energy
Borzoi	Sight Hound	very low	very low	moderately high	very low	very low
Boston terrier	Vermin Hunter	very low	moderately low	moderately low	moderately high	very low
Bouvier des Flandres	Drover	moderately high	very high	very low	moderately high	very low
Boxer	Personal Protection	very high	moderately low	moderately low	moderately high	moderately low
Briard	Drover	very high	moderately low	very high	moderately low	very high
Brittany	Pointer	moderately low	moderately high	very high	very high	moderately high
Brussels griffon	Companion	moderately low	moderately high	moderately high	moderately high	very high
Bull terrier	Fighting Dog	very high	very low	moderately low	moderately low	moderately low
Bulldog	Companion	very low	very low	very low	very high	moderately high
Bullmastiff	Guard	very high	moderately low	very low	moderately low	very low
Cairn terrier	Vermin Hunter	very high	moderately high	moderately low	moderately high	moderately high
Cardigan corgi	Drover	moderately low	very high	moderately low	moderately low	moderately high
Cavalier King Charles spaniel	Companion	very low	moderately low	very low	very high	very low
Chesapeake Bay retriever	Retriever	moderately high	moderately low	moderately high	moderately high	moderately low
Chihuahua	Companion	moderately low	moderately low	moderately high	very low	very high
Chow Chow	Spitz	very high	very low	very low	very low	very low
Clumber spaniel	Spaniel	very low	very low	very low	very high	very low
Cocker spaniel	Spaniel	very low	very high	moderately high	very high	moderately low
Collie	Herding	moderately low	very high	moderately high	moderately high	very low
Curly Coat retriever	Retriever	moderately low	moderately high	very high	moderately low	very low
Dachshund	Vermin Hunter	moderately high	very low	very high	very low	very high
Dalmatian	Guard	very high	very low	very high	moderately low	very high
Dandie Dinmont terrier	Vermin Hunter	moderately low	moderately high	very low	moderately low	moderately low
Doberman Pinscher	Personal Protection	moderately high	very high	very high	moderately low	moderately high
English cocker spaniel	Spaniel	very low	moderately high	moderately low	very high	moderately low

Breed	Group/ Function	Dominance/ Territoriality	Intelligence/ Learning Ability	Emotional Reactivity	Sociability	Energy
English foxhound	Scent Hound	moderately low	moderately low	very high	moderately high	moderately high
English springer spaniel	Spaniel	moderately low	moderately high	very high	very high	moderately high
English toy spaniel	Companion	very low	moderately low	very low	moderately low	very low
Field spaniel	Spaniel	moderately high	very high	very low	very high	moderately high
Flat coat retriever	Retriever	moderately low	moderately high	very high	very high	moderately high
French bulldog	Companion	moderately low	moderately low	very low	very low	moderately low
German shepherd	Herding	very high	very high	moderately low	moderately high	moderately high
German short-haired pointer	Pointer	very high	very high	very high	moderately high	very high
German wire-haired pointer	Pointer	very high	moderately high	moderately high	moderately low	very high
Giant schnauzer	Personal Protection	very high	very high	moderately low	very low	moderately high
Golden retriever	Retriever	very low	very high	moderately low	very high	moderately low
Gordon setter	Setter	very high	moderately high	moderately high	moderately high	moderately high
Great Dane	Guard	moderately low	very low	very low	very high	very low
Great Pyrenees	Guard	moderately high	moderately low	very low	moderately high	very low
Greyhound	Sight Hound	very low	very low	very high	very low	moderately low
Harrier	Scent Hound	very low	moderately high	moderately high	moderately high	moderately low
Irish setter	Setter	moderately low	moderately low	very high	very high	moderately high
Irish terrier	Vermin Hunter	very high	moderately high	moderately high	very low	moderately high
Irish water spaniel	Spaniel	moderately low	moderately high	very high	very low	moderately high
Irish wolfhound	Sight Hound	very low	moderately low	very low	moderately low	very low
Italian greyhound	Companion	very low	very low	very high	moderately low	very low
Japanese chin	Companion	very low	very high	very low	moderately high	very low
Keeshond	Spitz	very low	moderately high	moderately high	very high	moderately low
Kerry blue terrier	Vermin Hunter	very high	very high	moderately low	moderately high	moderately high

Breed	Group/ Function	Dominance/ Territoriality	Intelligence/ Learning Ability	Emotional Reactivity	Sociability	Energy
Komondor	Guard	very high	moderately low	moderately high	very low	moderately high
Kuvasz	Guard	very high	moderately high	moderately high	very low	moderately high
Labrador retriever	Retriever	moderately high	very high	very low	very high	very low
Lakeland terrier	Vermin Hunter	moderately low	moderately low	moderately low	moderately low	very high
Lhasa apso	Companion	moderately low	very low	moderately high	very low	moderately high
Maltese	Companion	very low	very low	moderately low	moderately low	moderately high
Manchester terrier	Vermin Hunter	moderately low	moderately high	very high	very low	very high
Mastiff	Guard	moderately high	very low	very low	moderately high	very low
Miniature pinscher	Companion	moderately high	moderately low	very high	very low	very high
Miniature poodle	Retriever	moderately low	very high	moderately high	moderately high	moderately high
Miniature schnauzer	Vermin Hunter	very high	moderately high	moderately low	moderately low	very high
Newfoundland	Draft	very low	moderately high	very low	very high	very low
Norfolk terrier	Vermin Hunter	moderately high	moderately high	moderately low	moderately low	very high
Norwegian elkhound	Spitz	moderately high	very high	very low	moderately high	moderately low
Norwich terrier	Vermin Hunter	moderately high	moderately high	moderately low	moderately low	very high
Nova Scotia duck tolling retriever	Retriever	moderately low	moderately high	very high	very high	moderately high
Old English sheepdog	Herding	moderately high	very low	moderately high	very high	moderately low
Otterhound	Scent Hound	moderately low	very low	moderately high	very high	very low
Papillon	Companion	very low	very high	moderately high	moderately high	moderately low
Parson Jack Russell terrier	Vermin Hunter	very high	very low	very high	moderately high	moderately high
Pekingese	Companion	moderately low	very low	very low	very low	very low
Pembroke corgi	Drover	moderately low	very high	moderately low	moderately low	moderately high
Pointer	Pointer	moderately high	very low	very high	moderately high	very high
Pomeranian	Companion	moderately low	very low	moderately low	very low	moderately high

Breed	Group/ Function	Dominance/ Territoriality	Intelligence/ Learning Ability	Emotional Reactivity	Sociability	Energy
Portuguese water dog	Multipurpose Sporting	moderately high	very high	very high	moderately low	moderately high
Pug	Companion	very low	moderately low	very low	moderately high	moderately high
Puli	Herding	very high	very high	very high	very low	very high
Rhodesian ridgeback	Personal Protection	very high	moderately low	moderately low	very low	moderately low
Rottweiler	Guard	very high	moderately high	moderately low	moderately high	very low
Saint Bernard	Draft	moderately low	very low	very low	moderately low	very low
Saluki	Sight Hound	very low	very low	very high	very low	moderately low
Samoyed	Spitz	moderately high	very low	very high	moderately low	very high
Schipperke	Spitz	moderately high	very high	very high	very low	moderately high
Scottish deerhound	Sight Hound	very low	very low	very low	very high	very low
Scottish terrier	Vermin Hunter	very high	very low	moderately high	very low	very high
Sealyham terrier	Vermin Hunter	moderately high	moderately low	very low	moderately low	very low
Shetland sheepdog	Herding	very low	very high	very high	moderately low	very high
Shih Tzu	Companion	very low	moderately low	very low	moderately high	moderately low
Siberian husky	Spitz	moderately high	very low	moderately high	very high	very high
Silky terrier	Vermin Hunter	moderately high	moderately high	moderately low	moderately high	moderately high
Skye terrier	Vermin Hunter	very low	moderately low	moderately low	very low	very low
Smooth fox terrier	Vermin Hunter	very high	moderately low	moderately high	moderately high	very high
Soft-coated Wheaten terrier	Vermin Hunter	moderately high	moderately high	moderately low	very high	moderately high
Staffordshire bull terrier	Fighting Dog	very high	moderately low	very low	moderately low	moderately high
Standard poodle	Retriever	moderately high	very high	moderately low	moderately high	very low
Standard schnauzer	Personal Protection	very high	very high	moderately low	moderately low	very high
Sussex spaniel	Spaniel	very high	very low	very low	very high	very low
Tibetan spaniel	Companion	moderately low	very low	very low	very low	very low
Tibetan terrier	Companion	moderately high	moderately low	moderately low	moderately low	moderately low
Toy poodle	Companion	very low	very high	very high	very low	very high

Breed	Group/ Function	Dominance/ Territoriality	Intelligence/ Learning Ability	Emotional Reactivity	Sociability	Energy
Vizsla	Multipurpose Sporting	very low	moderately high	moderately high	very high	very low
Weimaraner	Multipurpose Sporting	very high	moderately low	moderately high	moderately high	very high
Welsh springer spaniel	Spaniel	moderately high	moderately high	moderately high	moderately high	moderately high
Welsh terrier	Vermin Hunter	moderately high	moderately low	moderately high	moderately low	very high
West Highland white terrier	Vermin Hunter	very high	very low	moderately high	moderately high	very high
Whippet	Sight Hound	very low	moderately low	moderately high	very low	moderately low
Wire fox terrier	Vermin Hunter	very high	very low	very high	moderately high	very high
Wire haired pointing griffon	Pointer	moderately high	moderately high	very high	moderately low	moderately high
Yorkshire terrier	Companion	moderately high	moderately low	very high	moderately low	very high

SELECTED BIBLIOGRAPHY

The following contains a partial list of bibliographic references for the material in the book. I have tried to include review material whenever possible so that the reader can use these citations as a starting point for finding the more specific research that they may be interested in. If a reprint or a later edition of a work was consulted, that source, rather the original is listed.

Adams, B., Chan, A., Callahan, H., and Milgram, N. W. 2000. The canine as a model of human cognitive aging: recent developments. *Progress in Neuro-Psychopharmacology & Biological Psychiatry* 24, 675–92.

Alanen, L. 2003. *Descartes's concept of mind.* Cambridge, MA: Harvard University Press.

Bartlett, M. 1979, March. A novice looks at puppy aptitude testing. *American Kennel Club Gazette* 31–42.

———. 1979, March. A novice looks at puppy testing. *Purebred Dogs/American Kennel Gazette* 31–42.

———. 1985, March. Puppy aptitude testing: A new look. *Purebred Dogs/American Kennel Gazette* 30–35, 64.

Bavidge, M. 1994. *Can we understand animal minds?* London: Bristol Classical Press.

Bekoff, M., Allen, C., and Burghardt, G. M. 2002. *The cognitive animal: Empirical and theoretical perspectives on animal cognition.* Cambridge, MA: MIT Press.

Black, J. E., Isaacs, K. R., Anderson, B. J., Alcantara, A. A., and Greenough, W. T. 1990. Learning causes synptogenesis, whereas motor activity causes angiogenesis, in cerebellar cortex of adult rats. *Proceedings of the National Academy of Science of the United States of America* 87, 5568–5572.

Brenoe, U. T., Larsgard, A. G., Johannessen, K.P., and Uldal, S. H. 2002. Estimates of genetic parameters for hunting performance traits in three breeds of gun hunting dogs in Norway. *Applied Animal Behavior Science* 77, 209–215.

Brody, N. 1998. *Personality psychology: The science of individuality.* Upper Saddle River, NJ: Prentice Hall.

Cairns, R. B., and Weboff, J. 1967. Behavior development in the dog: An interspecific analysis. *Science* 1958, 1070–72.

Campbell, W. E. 1972. A behavior test for puppy selection. *Modern Veterinary Practice* 12, 29–33.

———. 1975. *Behavior problems in dogs.* Santa Barbara: American Veterinary Publications.

Cartwright, J. 2002. *Determinants of animal behaviour.* New York: Routledge.

Chan, A. D., Nippak, P. M., Murphey, H., Ikeda-Douglas, C. J., Muggenburg, B., Head, E., Cotman, C. W., and Milgram, N. W. 2002. Visuospatial impairments in aged canines *(Canis familiaris)*: The role of cognitive-behavioral flexibility. *Behavioral Neuroscience* 116, 443–54.

Chance, P. 2003. *Learning and behavior.* Belmont CA: Wadsworth.

Clarke, C. H. D. 1971, April. The beast of Gévaudan. *Natural History* 70–72.

Clarke, R. S., Heron, W., and Fetherstonhaugh, M. L. 1951. Individual differences in dogs: Preliminary report on the effects of early experience. *Canadian Journal of Psychology* 5, 150–156.

Cohn, J. 1997. How wild wolves became domestic dogs. *Bioscience* 47, 725–729.

Cooper, J. J., Ashton, C., Bishop, S., West, R., Mills, D. S., and Young, R. J. 2003. Clever hounds: social cognition in the domestic dog *(Canis familiaris). Applied Animal Behavior Science* 81, 229–224.

Coppinger, R., and Coppinger, L. 2001. *Dogs: A startling new understanding of canine origin, behavior and evolution.* New York: Scribner.

Coren, S. 1998. *Why we love the dogs we do.* New York: Free Press.

———. 2000. *How to speak dog: Mastering the art of dog-human communication.* New York: Free Press.

———. 2002. *The pawprints of history: Dogs and the course of human events.* New York: Free Press.

———. 2004. *How dogs think: Understanding the canine mind.* New York: Free Press, pp. i–xiv, 1–351.

———. 2006. *The intelligence of dogs: Canine consciousness and capabilities.* New York: Free Press.

Crist, E. 1999. *Images of animals: Anthropomorphism and animal mind.* Philadelphia: Temple University Press.

Csányi, V. 2005. *If dogs could talk.* New York: North Point Press.

Darwin, C. 1890. *The descent of man.* London: J. Murray.

———. 1964. *On the origin of species.* Cambridge, MA: Harvard University Press.

———. 1998. *The expression of the emotions in man and animals.* 3rd ed., with an introduction, afterword, and commentaries by Paul Ekman. London: HarperCollins.

Denny, M. R. 1991. *Fear, avoidance, and phobias: A fundamental analysis.* Hillsdale, NJ: Erlbaum.

Diamond, M. C. 1988. *Enriching heredity.* Free Press, New York.

Dodman, N. H. 2002. *If only they could speak.* New York: WW Norton.

Dore, F. Y., and Dumas, C. 1987. Psychology of animal cognition: Piagetian studies. *Psychological Bulletin* 102, 219–233.

Feddersen-Petersen, D. U. 2001. Biology of aggression in domestic dogs. *Deutsche Tieraerztliche Wochenschrift* 108, 94–101.

Federoff, N. E., and Nowak, R. M. 1997. Man and his dog. *Science* 278: 205.

Fentress, J. C. 1983. A view of ontogeny. In Eisenberg, J. F., and Kleiman, D. G., eds. *Advances in the study of mammalian behavior.* Lawrence, KS: the American Society of Mammalogists, 24–64.

———. 1992. The covalent animal: On bonds and their boundaries in behavioral research, in Davis, H., and Balfour, D., eds. *The inevitable bond: Examining scientist-animal interactions.* Cambridge: Cambridge University Press; 44–71.

Fox, M. W. 1971. *Integrative development of brain and behavior in the dog.* Chicago: University of Chicago Press.

Fox, M. 1980. *The soul of the wolf.* Boston: Little, Brown.

Frank, H. (ed.). 1987. *Man and wolf: Advances, issues, and problems in captive wolf research.* Perspectives in Vertebrate Science, no. 4. Dordrecht, Netherlands: Dr. W. Junk Publishers.

Frieman, J. 2002. *Learning and adaptive behavior.* Belmont, CA: Wadsworth.

Gershman, K. A., Sacks, J. J., and Wright, J. C. 1994. Which dogs bite? A case-control study of risk factors. *Pediatrics* 93, 913–917.

Gloyd, J. 1992. Wolf hybrids—A biological time bomb? *Journal of the American Veterinary Medical Association* 201, 381–382.

Goddard, M. E., and Beilharz, R. G. 1983. Genetics of traits which determine the suitability of dogs as guide-dogs for the blind. *Applied Animal Ethology* 9, 299–315.

———. 1985. A multivariate analysis of the genetics of fearfulness in potential guide dogs. *Behavior Genetics* 15, 69–89.

Goodwin, D., Bradshaw, J. W. S., and Wickens, S. M. 1997. Paedomorphosis affects agonistic visual signals of domestic dogs. *Animal Behavior* 53, 297, 304.

Gosling, S. D., and John, O. P. 1999. Personality dimensions in nonhuman animals: A cross-species review. *Current Directions in Psychological Science* 8, 69–75.

Gosling, S.D., and Kwan, V. S. Y. 2003. A dog's got personality: A cross-species comparative approach to personality judgments in dogs and humans. *Journal of Personality and Social Psychology* 85, 1161–1169.

Hare, B., Brown, M., Williamson, C., and Tomasello, M. 2002. The domestication of social cognition in dogs. *Science* 298, 1634–1636.

Hare, B., and Wrangham, R. 2002. Integrating two evolutionary models for the study of social cognition. In Bekoff, M., and Allen, C. eds. *The cognitive animal: Empirical and theoretical perspectives on animal cognition.* Cambridge, MA: MIT Press, 363–369.

Hart, B. L., and Hart, L. A. 1985. Selecting pet dogs on the basis of cluster analysis of breed behavior profiles and gender. *Journal of the American Veterinary Medical Association* 186, 1181–85.

———. 1988. *The perfect puppy.* New York: Freeman.

Hebb, D. O. 1946. The objective description of temperament. *American Psychologist* 1, 275–276.

———. 1947. The effects of early experience on problem solving at maturity. *American Psychologist* 2, 306–307.

———. 1949. Temperament in chimpanzees: I. Method of analysis. *Journal of Comparative and Physiological Psychology* 42, 192–206.

———. 1949. *The organization of behavior: a neuropsychological theory.* Oxford, England: Wiley.

Hepper, P. G., and Cleland, J. 1998–1999. Developmental aspects of kin recognition. *Genetica* 104, 199–205.

Hsu, Y., and Serpell, J. A. 2003. Development and evaluation of a novel method for evaluating behavior and temperament in pet dogs. *Journal of the American Veterinary Medical Association* 223, 1293–1300.

Humphrey, E., and Warner, L. 1974. *Working dogs.* Palo Alto: National Press.

Igel, G. J., and Calvin, A. D. 1960. The development of affectional responses in infant dogs. *Journal of Comparative and Physiological Psychology* 53, 302–305.

Jagoe, A., and Serpell, J. 1996. Owner characteristics and interactions and the prevalence of canine behaviour problems. *Applied Animal Behavior Science* 47, 31–42.

Katcher, A. H., and Beck, A. M. 1986. Dialogue with animals. *Transactions & Studies of the College of Physicians of Philadelphia* 8, 105–12.

Kempermann, G., Kuhn, H. G., and Gage, F. H. 1997. More hippocampal neurons in adult mice living in an enriched environment. *Nature* 386, 493–495.

Kirkness, E. F., Bafna, V., Halpern, A. et al., 2003. The dog genome: Survey sequencing and comparative analysis. *Science* 301, 1898–1903.

Klinghammer, E., and Goodmann, P. A. 1987. Socialization and management of wolves in captivity. In Frank, H., ed. *Man and wolf.* Dordrecht: Dr. W. Junk Publishers, 31–59.

Klintsova, A., and Greenough, W. T. 1999. Synaptic plasticity in cortical systems. *Current Opinion in Neurobiology* 9, 203–208.

Krech, D., Rosenzweig, M. R., and Bennett, E. L. 1960. Effects of environmental complexity and training on brain chemistry. *Journal of Comparative and Physiological Psychology* 53, 509–519.

Kroll, T. L., Houpt, K. A., and Erb, H. N. 2004. The use of novel stimuli as indicators of aggressive behavior in dogs. *Journal of the American Animal Hospital Association* 40, 13–19.

Leonard, J. A., Wayne, R. K., Wheeler, J., Valadez, R., Guillen, S., and Vila, C. 2002. Ancient DNA evidence for old world origin of new world dogs. *Science* 298: 1613–1616.

Lindsay, S. R. 2000. *Handbook of applied dog behavior and training* (vols. 1 and 2). Ames, IA: Iowa State Press.

Lockwood, R. 1992, fall. Dangerous dogs revisited. *Humane Society,* U.S. News, 20–22.

Lockwood, R., and Rindy, K. 1987. Are "pit-bulls" different? An analysis of the pit bull terrier controversy. *Anthrozoös* 1(1): 2–8

Lorenz, Konrad. 1952. *King Solomon's ring: new light on animal ways.* New York: Crowell.

———. 1965. *Evolution and modification of behavior.* Chicago: University of Chicago Press.

McCrae, R. R., and Costa, P. T., Jr. 1987. Validation of the five-factor model of personality across instruments and observers. *Journal of Personality and Social Psychology* 52, 81–90.

Mech, L. D. 1981. *The wolf: Ecology and behavior of an endangered species.* Minneapolis: University of Minnesota Press.

Mech, L. D. et al. 1998. *The wolves of Denali.* Minneapolis: University of Minnesota Press.

Milgram, N. W., Head, E., Muggenburg, B., Holowachuk, D., Murphey, H., Estrada, J., Ikeda-Douglas, C. J., Zicker, S. C., and Cotman, C. W. 2002. Landmark discrimination learning in the dog: effects of age, an antioxidant fortified food, and cognitive strategy. *Neuroscience & Biobehavioral Reviews* 26, 679–95.

Mitchell, R. W., Thompson, N. S., and Miles, H. L. 1997. *Anthropomorphism, anecdotes, and animals.* Albany: State University of New York Press.

Murphy, J. A. 1995. Assessment of the temperament of potential guide dogs. *Anthrozoös* 8, 224–228.

Neff, M. W., Broman, K. W., Mellersh, C. S. et al. 1999. A second-generation genetic linkage map of the domestic dog, Canis familiaris. *Genetics* 151, 803–820.

Netto, W. J., and Planta, D. J. U. 1997. Behavioural testing for aggression in the domestic dog. *Applied Animal Behavior Science* 52, 243–263.

Olsen, S. J., and J. Olsen, J. W. 1977. The Chinese wolf, ancestor of new world dog. *Science* 197:533–535.

Overall, K. L. 2000. Natural animal models of human psychiatric conditions: Assessment of mechanism and validity. *Progress in Neuro-Psychopharmacology & Biological Psychiatry* 24, 727–776.

———. 2006. Behavioral Genetics II: Genetics of neurochemistry of normal and abnormal aggression. *Proceedings of "The dog and its mind: Revolutionary insights,"* Professional Animal Behavior Associates meeting, Guelph, Ontario.

Pearce, J. M. 1997. *Animal learning and cognition: an introduction.* Hove, East Sussex: Psychology Press.

Pennisi, E. 2005. How did cooperative behavior evolve? *Science* 309, 93–96.

Pfaffenberger, C. J. 1963. *The new knowledge of dog behavior.* New York: Howell Book House.

Pfaffenberger, C. J., Scott, J. P., Fuller, J. L., Ginsburg, B. E., and Bielfelt, S. W. 1976. *Guide dogs for the blind: Their selection, development, and training.* Amsterdam: Elsevier Scientific.

Podberscek, A. L. and Serpell, J. 1996. The English cocker spaniel: Preliminary findings on aggressive behavior. *Applied Animal Behavior Science* 47, 75–89.

Pongrácz, P., Miklósi, A., Kubinyi, E., Topául J., and Csányi, V. 2003. Interaction between individual experience and social learning in dogs. *Animal Behavior* 65, 595–603.

Rampon, C., Jiang, C. H., Dong, H., Tang, Y.-P., Lockhart, D. J., Schultz, P. G., Tsien, J. Z., Hu, Y. 2000. Effects of environmental enrichment on gene expression in the brain. *Proceedings of the National Academy of Science of the United States of America* 97, 12880–12884.

Reisner, I. R. 2003. Differential diagnosis and management of human-directed aggression in dogs. *Veterinary Clinics of North America—Small Animal Practice* 33, 303–320.

Rosenzweig, M. R., and Bennett, E. L. 1996. Psychobiology of plasticity: Effects of training and experience on brain and behavior. *Behavior and Brain Research* 78, 57–65.

Rothwell, J. H., and Weatherwax, R. B. 1950. *The story of Lassie—His discovery and training.* London: W. H. Allen.

Sablin, and Khlopachev, G. A. 2002. The earliest ice age dogs: Evidence from Eliseevichi 1. *Current Anthropology* 43, 795–799.

Sacks, J. J., Sinclair, L., Gilchrist, J., Golab, G. C., Lockwood, R. 2000. Breeds of dogs involved in fatal human attacks in the United States between 1979 and 1998. *Journal of the American Veterinary Medical Association* 217, 836–840.

Scott, J. P. 1967. The development of social motivation. In *Nebraska symposium on motivation.* Lincoln, NE: University of Nebraska Press, 111–132.

———. 1958. Critical periods in the development of social behavior in puppies. *Psychosomatic Medicine* 20, 42–54.

———. 1977. Social genetics. *Behavior Genetics* 7, 327–346.

Scott, J. P., Stewart, J. M., and de Ghett, V. J. 1974. Critical periods in the organization of systems. *Developmental Psychobiology* 7, 489–513.

Scott, J. P., and Fuller, J. L. 1965. *Genetics and the social behavior of the dog.* Chicago: University of Chicago Press.

Serpell, J., and Jagoe, J. A. 1995. Early experience and the development of behavior. In Serpell, J., ed. *The domestic dog: Its evolution, behaviour and interactions with people.* New York: Cambridge University Press, 79–102.

Serpell, J. A. 1995. *The domestic dog: its evolution, behaviour and interactions with people.* Cambridge, England: Cambridge University.

Serpell, J. A., and Hsu, Y. 2001. Development and evaluation of a novel method for evaluating behavior and temperament in guide dogs. *Applied Animal Behavior Science* 72, 347–364.

Sheppard, G., and Mills, D. S. 2002. The development of a psychometric scale for the evaluation of the emotional predispositions of pet dogs. *International Journal of Comparative Psychology* 15, 201–222.

Stevens, J. R., Cushman, F. A., and Hauser, M. D. 2005. Evolving the Psychological mechanisms for Cooperation. *Annual Review of Ecology, Evolution, & Systematics* 36, 499–518.

Svartberg, K. 2002. Shyness-boldness predicts performance in working dogs. *Applied Animal Behavior Science* 79, 157–174.

Svartberg, K., and Forkman, B. 2002. Personality traits in the domestic dog (*Canis familiaris*). *Applied Animal Behavior Science* 79, 133–155.

Takeuchi, Y., and Houpt, K. 2003. Behavior genetics. *Veterinary Clinics of North America—Small Animal Practice* 33, 345–363.

Trut, L. N. 1999. Early canid domestication: The farm fox experiment. *American Scientist* 87: 160–169.

Van der Borg, J., Netto, W., and Planta, D. 1991. Behavioral testing of dogs in animal shelters to predict problem behavior. *Applied Animal Behavior Science* 32, 237–251.

Vila, C., Maldonado, J. E., and Wayne, R. K. 1999. Phylogenetic relationships, evolution, and genetic diversity of the domestic dog. *The Journal of Heredity* 90(1): 71–77.

Weiss, E. 2002. Selecting shelter dogs for service dog training. *Journal of Applied Animal Welfare Science* 5, 143–162.

West, R., and Young, R. J. 2002. Do domestic dogs show any evidence of being able to count? *Animal Cognition* 5, 183–186.

Wheeler, S. J. 1995. *BSAVA manual of small animal neurology.* Cheltenham, Gloucestershire, UK: British Small Animal Veterinary Association.

Wiggins, J. S. 1996. *The five-factor model of personality: Theoretical perspectives.* New York: Guilford Press.

Wilkins, A. S. 1993. *Genetic analysis of animal development.* New York: Wiley-Liss.

Wilsson, E. 1984–1985. The social interaction between mother and offspring during weaning in German shepherd dogs: Individual differences between mothers and their effects on offspring. *Applied Animal Behavior Science* 13, 101–112.

Wilsson, E., and Sundgren, P. 1997. The use of a behavior test for selection of dogs for service and breeding: II. Heritability for tested parameters and effect of selection based on service dog characteristics. *Applied Animal Behavior Science* 54, 235–241.

———. 1997. The use of a behavior test for the selection of dogs for service and breeding: I. Method of testing and evaluating test results in the adult dog, demands on different kinds of service dogs, sex and breed differences. *Applied Animal Behavior Science* 53, 279–295.

———. 1998. Behavior test for eight-week old puppies: Heritabilities of tested behavior traits and its correspondence to later behavior. *Applied Animal Behavior Science* 58, 151–162.

——— 1998. Effects of weight, litter size and parity of mother on the behavior of the puppy and the adult dog. *Applied Animal Behavior Science* 56, 245–254.

Woolpy, J. H. 1968. Socialization of wolves. In *Animal and human: Scientific proceedings of the American Academy of Psychoanalysis; Science and Psychoanalysis* vol. 12, J. H. Masserman, ed. New York: Grune and Stratton, 82–94.

———. 1968. The social organization of wolves. *Natural History* 77(5):46–53.

Zimen, E. 1987. Ontogeny of approach and flight behavior towards humans in wolves, poodles and wolf-poodle hybrids. In Frank, H., ed. *Man and wolf.* Dordrecht, Netherlands: Dr. W. Junk Publishers, 275–292.

———. 1981. *The wolf: A species in danger.* New York: Delacorte Press.

Zotterman, Y. 1967. *Sensory mechanisms.* New York: Elsevier.

INDEX

ABOUT THE AUTHOR

Stanley Coren, PhD, FRSC, is a professor of psychology at the University of British Columbia and a recognized expert on dog-human interaction. He has appeared on *Dateline*, the *Oprah Winfrey Show, Good Morning America*, and National Public Radio, and hosts a weekly television show, *Good Dog!*, currently showing nationally in Canada, Australia, and New Zealand. He lives in Vancouver, British Columbia, with his wife and her cat, in addition to a beagle, a Cavalier King Charles spaniel, and a Nova Scotia duck tolling retriever.